The Metabotropic Glutamate Receptors

The Receptors

Series Editor

David B. Bylund
University of Nebraska Medical Center, Omaha, NE

Board of Editors

The Metabotropic Glutamate Receptors

Edited by

P. Jeffrey Conn
Emory University, Atlanta, GA

Jitendra Patel
Zeneca Pharmaceuticals Group, Wilmington, DE

Humana Press
Totowa, New Jersey

QP
364
.7
.M466
1994

© 1994 Humana Press Inc.
999 Riverview Drive, Suite 208
Totowa, New Jersey 07512

This publication is printed on acid-free paper. ∞
ANSI Z39.48-1984 (American National Standards Institute) Permanence of Paper
for Printed Library Materials.

Photocopy Authorization Policy:
Authorization to photocopy items for internal or personal use, or the internal or personal use
of specific clients, is granted by Humana Press Inc., **provided** that the base fee of US $3.00
per copy, plus US $00.20 per page, is paid directly to the Copyright Clearance Center at 222
Rosewood Drive, Danvers, MA 01923. For those organizations that have been granted a
photocopy license from the CCC, a separate system of payment has been arranged and is
acceptable to Humana Press Inc. The fee code for users of the Transactional Reporting
Service is: [0-89603-291-4/94 $3.00 + $00.20].

Printed in the United States of America. 10 9 8 7 6 5 4 3 2 1

Library of Congress Cataloging in Publication Data

The Metabotropic glutamate receptors / edited by P. Jeffrey Conn and
    Jitendra Patel.
        p. cm. -- (The Receptors)
    Includes index.
    ISBN 0-89603-291-4
    1. Glutamate acid--Receptors. I. Conn, P. Jeffrey. II. Patel,
Jitendra. III. Series.
    [DNLM: 1. Receptors, Metabotropic Glutamate--metabolism.
2. Receptors, Metabotropic Glutamate--physiology. 3. Second
Messenger Systems--physiology. QU 60 M588 1994]
QP364.7.M466 1994
612'.01575--dc20
DNLM/DLC
for Library of Congress
                                                        94-18190
                                                        CIP

Preface

It is the goal of *The Metabotropic Glutamate Receptors* to provide a comprehensive and forward-thinking review of the tremendous advances that have occurred in less than a decade of metabotropic glutamate receptors (mGluR) research. Virtually every area of mGluR research is covered, including the molecular biology, pharmacology, anatomical distribution, and physiological and pathological roles of mGluRs. It is our intention that this volume not only summarize what is now known about the mGluRs, but also illuminate the areas in which there is the greatest need for focused research.

Glutamic acid is an amino acid that has long been known to play several important metabolic roles in central and peripheral tissues and to be a component of several naturally occurring molecules. The first evidence that glutamate may also serve as a neurotransmitter in the central nervous system (CNS) came in the late 1950s and early 1960s when glutamate and other acidic amino acids were found to induce behavioral convulsions when topically applied to the cortex and to excite a wide variety of central neurons. These findings spurred a massive research effort that quickly established glutamate as the primary excitatory neurotransmitter in the vertebrate CNS. One of the most striking characteristics of glutamate that was quickly recognized was its ubiquitous role in serving as the neurotransmitter at the vast majority of excitatory synapses in the brain. It is now clear that most central neuronal circuits involve glutamatergic neurotransmission at some level.

Although early studies suggested that most of glutamate's actions as a neurotransmitter were mediated by the activation of ligand-gated cation channels, a few papers began to appear in the early 1980s that suggested that glutamate and related excitatory amino acids have some actions that could not be explained by direct opening of glutamate-gated channels. However, it was not until 1985–1987 that the first unequivocal evidence for the existence of G-protein-linked glutamate receptors, or metabotropic glutamate receptors, began to appear. Since that time, it has become clear that a large family of mGluRs exist. Members of the mGluR family are heterogeneously distributed in virtually every region of the mammalian brain, where they are coupled to a wide variety of second messenger systems.

The discovery of mGluRs dramatically alters the traditional view of glutamatergic neurotransmission. Since activation of mGluRs can modulate activity in glutamatergic circuits in a manner previously associated only with neuromodulators, the mGluRs provide a mechanism by which glutamate can modulate or fine tune activity at the same synapses at which it elicits fast synaptic responses. Because of the ubiquitous distribution of glutamatergic synapses, mGluRs have the potential of participating in virtually all known functions of the CNS. Indeed, it the short history of mGluR research, these receptors have already been implicated in a variety of processes including motor control, learning and memory, developmental plasticity, vision, sensory processing, nociception, regulation of cardiovascular function, epileptogenesis, and responses to neuronal injury. In addition, because of the wide diversity and heterogeneous distribution of mGluR subtypes, the opportunity exists to develop pharmacological agents that selectively interact with mGluRs involved in only one or a limited number of CNS functions. Thus, gaining a detailed understanding the specific roles of mGluRs might well have a dramatic impact on the development of novel treatment strategies for a variety of neurological disorders.

We would like to thank our many colleagues in the mGluR field for stimulating discussions that have helped shape the direction of this book and for sharing data prior to its publication that has been discussed in various chapters. We especially thank Dr. Darryle Schoepp, who had provided valuable insight and advice at all editorial stages of the book's production. Special thanks also go to Dr. Ferdinando Nicolleti and others who were involved in organizing the First Annual Meeting of Metabotropic Glutamate Receptors. This meeting took place a only weeks before submission of the final manuscripts that make up this volume and had a tremendous impact on its content. Information gleaned from this meeting made the book considerably more current than it otherwise would have been.

P. Jeffrey Conn
Jitendra Patel

Contents

Contributors

E. ARONICA • *Department of Pharmacology, University of Catania, Italy*

G. BATTAGLIA • *Department of Pharmacology, University of Catania, Italy*

VALERIE BOSS • *Department of Pharmacology, Emory University, Atlanta, GA*

V. BRUNO • *Department of Pharmacology, University of Catania, Italy*

G. CASABONA • *Department of Pharmacology, University of Catania, Italy*

M. V. CATANIA • *Department of Pharmacology, University of Catania, Italy*

DOROTHY S. CHUNG • *Department of Pharmacology, Emory University, Atlanta, GA*

D. F. CONDORELLI • *Department of Biochemistry, University of Catania, Italy*

P. JEFFREY CONN • *Department of Pharmacology, Emory University, Atlanta, GA*

A. COPANI • *Department of Pharmacology, University of Catania, Italy*

BEAT H. GÄHWILER • *Brain Research Institute, University of Zurich, Switzerland*

JOEL P. GALLAGHER • *Department of Pharmacology and Toxicology, University of Texas, Galveston, TX*

A. A. GENAZZANI • *Department of Pharmacology, University of Catania, Italy*

URS GERBER • *Brain Research Institute, University of Zurich, Switzerland*

ROBERT W. GEREAU IV • *Department of Pharmacology, Emory University, Atlanta, GA*

STEVEN R. GLAUM • *Department of Pharmacology, University of Chicago, Chicago, IL*

PETER KRISTENSEN • *Departments of Pathology and Histochemistry, Novo-Nordisk, Malov, Denmark*

M. R. L'EPISCOPO • *Department of Pharmacology, University of Catania, Italy*

RICHARD J. MILLER • *Department of Pharmacology, University of Chicago, Chicago, IL*

EILEEN MULVIHILL • *Zymogenetics, Seattle, WA*

F. NICOLETTI • *Department of Pharmacology, University of Catania, Italy; Department of Experimental Medicine and Biochemical Sciences, University of Perugia, Italy*

JITENDRA PATEL • *Zeneca Pharmaceuticals Group, Wilmington, DE*

DARRYLE D. SCHOEPP • *Central Nervous System Research, Eli Lilly and Co., Indianapolis, IN*

PATRICIA SHINNICK-GALLAGHER • *Department of Pharmacology and Toxicology, University of Texas, Galveston, TX*

PETER D. SUZDAK • *Department of Receptor Neurochemistry, Novo-Nordisk, Malov, Denmark*

CLAUDIA M. TESTA • *Department of Neurology, Massachusetts General Hospital, Harvard Medical School, Boston, MA*

CHRISTIAN THOMSEN • *Department of Receptor Neurochemistry, Novo-Nordisk, Malov, Denmark*

DANNY G. WINDER • *Department of Pharmacology, Emory University, Atlanta, GA*

ANNE B. YOUNG • *Department of Neurology, Harvard Medical School, Massachusetts General Hospital, Boston, MA*

FANG ZHENG • *Department of Pharmacology and Toxicology, University of Texas, Galveston, TX*

WILLIAM C. ZINKAND • *CNS Pharmacology, ICI Americas, Wilmington, DE*

CHAPTER 1

Molecular Cloning, Expression, and Characterization of Metabotropic Glutamate Receptor Subtypes

Peter D. Suzdak, Christian Thomsen, Eileen Mulvihill, and Peter Kristensen

1. Introduction

The excitatory amino acid, L-glutamate, is a primary neurotransmitter in excitatory synaptic pathways in the central nervous system (for a review, *see* Monaghan et al., 1989; Headley and Grillner, 1990; Mayer and Miller 1990; Nakanishi, 1992). L-Glutamate-mediated neurotransmission is involved in numerous neuronal functions, and excess glutamatergic stimulation may also be involved in the etiology of stroke, epilepsy, and neurodegenerative disorders (Monaghan et al., 1989; Meldrum and Garthwaite, 1990). Receptors for L-glutamate can be classified into two distinct groups based on their signal transduction pathways: (1) ionotropic glutamate receptors, which when activated are directly coupled to the opening of cationic channels (MacDermott et al., 1986; Murphy et al., 1987)—ionotropic glu-

The Metabotropic Glutamate Receptors Eds.: P. J. Conn and J. Patel
© 1994 Humana Press Inc., Totowa, NJ

tamate receptors have been further defined by pharmacological and electrophysiological selectivity for *N*-methyl-D-aspartate (NMDA), α-amino-3-hydroxyl-5-methyl-1-isoxazole-4-propionic acid (AMPA), and kainic acid—and (2) metabotropic glutamate receptors (mGluRs), which are G-protein-coupled receptors (Monaghan et al., 1989; Nahorski and Potter, 1989; Mayer and Miller, 1990; Schoepp and Conn, 1993).

The first mammalian mGluR was discovered by Sladeczek et al. (1985), with the observation that glutamate and quisqualate stimulated phospholipase C in cultured striatal neurons through a novel type of excitatory amino acid receptor. Over the past several years, our understanding of the biochemical and physiological roles of mGluRs has increased dramatically, with the availability of more selective ligands, and the cloning and expression of the mGluR family. mGluRs have been shown to affect multiple aspects of neuronal function, including changes in the activities of phospholipase C, phospholipase D, adenylyl cyclase, calcium channels, and potassium channels, as well as excitatory postsynaptic potentials, long-term potentiation (LTP), and long-term depression (LTD) (for a review, *see* Schoepp and Conn, 1993). In addition, mGluR ligands have been shown in vitro and in vivo either to enhance, or inhibit glutamatergic neurotransmission.

With the original isolation of the mGluR1a cDNA by Houamed et al. (1991) and Masu et al. (1991), and the subsequent isolation of six additional mGluR subtypes, a new chapter in the history of mGluR research has been opened. The differences in the agonist/antagonist selectivity, second-messenger system coupling, and distribution of these subtypes may provide new insight into the diversity of mGluR effects, and in due course provide a means for the preparation of new and selective drugs working at individual members of the receptor family. The present chapter will discuss the cloning, expression pattern, and pharmacological characterization of the mGluR family.

2. Molecular Cloning of mGluRs

Using a cDNA expression library derived from rat cerebellum, Houamed et al. (1991) and Masu et al. (1991) independently reported the expression cloning of mGluR1a. RNA transcribed from pooled

Table 1
Characteristics of Members of the mGluR Family as Predicted
from the cDNA Clones[a]

	Number of amino acids	Rank order of agonist potency
Group 1		
mGluR1a	1199	
mGluR1b	907	
mGluR1c	897	Quisqualate > glutamate > ibotenate > 1S,3R-ACPD
mGluR5a	1171	
mGluR5b	1193	
Group 2		
mGluR2	872	L-CCG-I > glutamate = 1S,3R-ACPD > ibotenate
mGluR3	879	> quisqualate
Group 3		
mGluR4a	912	
mGluR4b	983	L-AP4 > glutamate >> 1S,3R-ACPD > quisqualate
mGluR6	871	> ibotenate
mGluR7	914	

[a]The subtypes of mGluRs can be divided into three groups based on their amino acid sequence homologies (*see* Table 2), their coupling to second-messenger systems, and their agonist selectivities.

clones was injected into oocytes that were monitored for changes in calcium-dependent chloride conductance produced by glutamate, quisqualate, or 1S,3R-ACPD using voltage-clamp recording (Houamed et al., 1991; Masu et al., 1991). Positive pools were subdivided until a single clone was identified. This cDNA was subsequently used for isolation of additional mGluR subtypes from rat brain and retinal cDNA libraries (Tanabe et al., 1992, 1993; Abe et al., 1992; Minakami et al., 1993; Nakajima et al., 1993; Westbrook et al., 1993; Okamoto et al., 1994). Currently, seven mGluR subtypes have been isolated (Table 1), together with additional splice variants. Based on sequence homology (Table 2) and pharmacology, mGluRs can be subdivided into three subfamilies as indicated in Table 1.

The amino acid sequences of all isolated members of the mGluR family show a number of similar features (Fig. 1). Hydrophobicity analysis predicts that the hydrophobic signal peptide is followed by

Table 2
Amino Acid Sequence Homologies Among Members
of the mGluR Family[a]

	mGluR1	mGluR2	mGluR3	mGluR4	mGluR5	mGluR6	mGluR7
mGluR1	100	46	44	43	61	40	38
mGluR2		100	70	43	43	46	45
mGluR3			100	47	42	45	43
mGluR4				100	40	69	69
mGluR5					100	39	38
mGluR6						100	67
mGluR7							100

[a]The values represent the percentage of amino acid homologies among the various subtypes of the mGluR family. The data are from Houamed et al., 1991; Masu et al., 1991; Nakanishi, 1992; Okamoto et al., 1994; Tanabe et al., 1992.

a large extracellular N-terminal domain (>500 amino acids), seven transmembrane regions, and finally a C-terminal region with a variable length. Interestingly, alignment of the members of the mGluR family demonstrates that the regions conserved between individual members of the family are evenly distributed throughout the proteins (Fig. 2). This is in contrast to other G-protein-coupled receptors, like the β-adrenergic or dopamine receptor families, where the transmembrane regions show a very high degree of conservation. The alignment also demonstrates the conservation of cysteines in the extracellular N-terminal domain, an important finding when considering this region as the putative ligand binding domain *(see* Section 3.1.). It is also important to note that the third intracellular loop, which has been shown to be important for the functional coupling in G-protein-coupled receptors (Franke et al., 1990; Luttrell et al., 1993; Maggio et al., 1993), is highly conserved between mGluR subtypes despite their coupling to different second-messenger systems. Recent experiments have indeed shown that in the mGluR family, the second intracellular loop together with parts of the C-terminal are the regions most important for the G-protein coupling *(see* Section 3.2.).

The diversity within the mGluR family is further increased by the existence of variants with regard to C-terminal amino acid sequence for several subtypes, most likely arising through alterna-

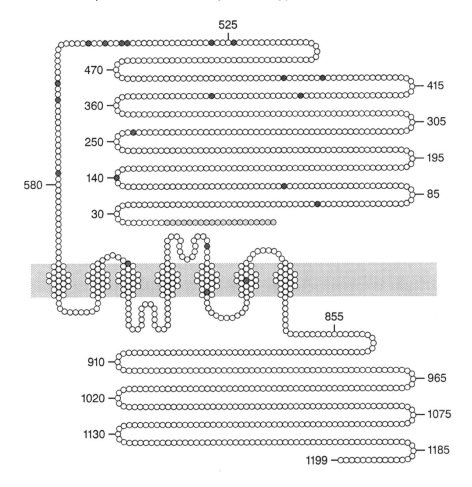

Fig. 1. Schematic representation of the structure of the mGluR. The cystein residues that are conserved are marked with filled circles. At the N-terminal domain is a stretch of 20 amino acids that serves as a signal peptide (light gray circles). From Houamed et al. (1991) and Masu et al. (1991).

tive splicing of pre-mRNA. The two alternatively spliced forms of mGluR1a have a shorter C-terminus (and thus resemble mGluR2–mGluR4a). The last 312 amino acids of mGluR1a are thus replaced by 20 new amino acids in mGluR1b (Tanabe et al., 1992; Pickering et al., 1993) and by 10 new amino acids in mGluR1c (Pin et al., 1992).

```
                    1                                                                                          110
mGluR1a   ........mvrllliffpmiFlemslpiMpdrkvllaqasqriSVaRMDGDVllGaLFsVHhqppaekvperkCGeIrEqyGlORvEAMFhrLDkiNaDPvLLPNITLGA
mGluR2    ........................mesllgtlalllL..wgavaegpakkvltLEGDIVLGGLFPVHqkGpae.....eCGpVnEhrGlORLEAMLFALDRINrDPhLLPgVrLGA
mGluR3    ........mkmmLtrLqiMLaiFskgFL...lslgdhnfmIrelkiEGDIVLGGLFPinekGtge.....eCGrInEDrGIORLEAMLFAiDeINkDnyLLPgvkLGV
mGluR4a   .msgkgqvawwwarlPlcLLislyqpWvpsigkpkghphmnsIRiDGDltLGGLFPVHqrGseqk.....aCGeikkeKGiRhLEAMLFALDRINrDPdLLPNITLGA
mGluR5a   .........mvllllsvlllkedvrgs.....aqsserrvVAhMpGDIiGaLFsVHhqptvdkvherkCGaVrEqyGlOrvEAMLhtLERiNsDPtLLPNITLGc
mGluR6    ..........MgrLPvlLL.......WLawwlsqagiacqagSV.RLaGqItLGGLFPVHArGaagr....aCGaikkEqGVhrLEAMLYALDRVNaDPeLLPgVrLGA
mGluR7    mvqigklirvltLmkFPccvLevLlcvLaaaargqemyaph..SI.RiEGDVtLGGLFPVHakGpsgv...pCGdikrEnGiHrLEAMLYALDqINsDPnLLPNTLGA

                    111                                                                                        220
mGluR1a   eIrDsCwhssvaALEQSiEFIRdSLisirDekDqlnRClpdgqtlppgrtkpIaqVIGpgsSSValQVqNLLqLFdiPQlaYsaTSiDLSDKtlYkYFlRVVPsDtLQAr
mGluR2    hILDsCSKDThALEQaLDFVRASL..srgaDgsthiCpdGsyathsdaPta.VtgVIGgSysdVSIQVANLLRLFqiPQISYASTsakLSDKsRYDYFARtVPPDfFOAr
mGluR3    hILDtCSRDTyALEQSlEFVRASL..tk.vDEaeymCpdGsyaigeniP1l.IagVIGgSySSVSIQVANLLRLFqiPQISYASTapDLSDnsRYDFFsRVVPsDtYQAq
mGluR4a   rILDTCSRDThALEQSLtFVRAIi...EKDgteVRGrrgppiitkPer.VVGVIGaSgSSVSIMVANiLRLFKiPQISYASTsmDLSDKiFkYFmRVVPsDaqOAr
mGluR5a   eIrDsCwhsavaALEQSiEFIRdSLis.sEeEEgiVRCvdGssf...rskkpIVGVIGpgsSValQVqNLLqLFniPQlaYsaTSmDLSDKtlYkYFmRVVPsDaqOAr
mGluR6    rILDTCSRDTyALEQaLsFVqAiirrgDgDEasVRCpgGvpPirsapPer.VVaVVGaSaSSVSImVANvLRLFqiPQISYASTapELSDstRYDFFsRVVPPDsYQAq
mGluR7    rILDTCSRDTyALEQSLtFVqAli....qkDtsdVRCrnGeppv.fvkPee.VVGVIGaSgSSVSImVANiLRLFqiPQISYASTapELSDdiRYDFFsRVVPPDsFQAq

                    221                                                                                        330
mGluR1a   AMlDIVkrYNMTYVSaVhtEGnYGESCmDAFKela.aqeGIClAhSdKI..ysnageksFDrllRKLrErlPKARVVvCFcegmtVRgLLsAmRRlgvvGeFsllGSDGW
mGluR2    AMaEIlRfFNMTYVSTVASEGdYGEtGIEAFeleaR.arnICVAtSeKVgRa..msraaFEgVVRaLLqk.PsARVaViFtrsEDmReLLaAtGrlN..asFtWaSDGW
mGluR3    AMaEIlRfFNMTYVSTVASEGdYGEtGIEAFeQeaR.lrnICIAtaeKVgRs..nirksYDsVIReLLqk.PNARVVVlFmrsDDsReLiaAnRvN..asFtWaSDGW
mGluR4a   AMVDIVRaLkWnYVSTlASEGsYGESRengGVCIAqsVrRepk..tgeFDkIIkrlLEt.sNARgIIiFaneDDIRrvLeAaRaNqtGhFFWmGSDsW
mGluR5a   AMVDIVkrYNMTYVSaVhtEGnYGESCmDAFKdmS.akeGIClAhSyKI..ysnageqsFDkllkklrshlPkARVVacFcegmtVRgLLmAmRRlglaGeFLllGSDGW
mGluR6    AMVDIVRaLgWnYVSTlASEGsYGESVEAFVQlSReagGVCIAqSiKIpRepk.pgeFhkVIRrLMEt.PNARgIIiFaneDDIRrvLeAtRqaNltGhFLWGSDsW
mGluR7    AMVDIVkaLgWnYVSTlASEGsYGEKgGVEsFtQiSkeagGICIAqsVrIpqerkdrtidFDrIIkgLLDt.PNsRaVViFandEDIkqiLaAakRadqvGhFLWVGSDsW

                    331                                                                                        440
mGluR1a   adrdeVIeGyEveAnGgITIKLqSpeVrsFDdYlKlrLDtNtRNPWFPEFWqhrFqCrLpghllenpnfkkvCTgnEs1..EenYvQDSKmgFVINAIYAMAHgLqnMH
mGluR2    GaIesVWaGsEraAeGAITIeLaSypIsdFasYFqSLdpwNNsRNPWFrEFWEErFhCsF......rqrd...Caah..slravpFEQESKImFVVNAVYAMAHALHnMH
mGluR3    GaqesIVKGsEhvAyGAITIeLaShpVrqFDrYFqSLnpyNNhRNPWFrDFWEqkFQCsL...qnkrnhrqvCdkhla.idsnYEQESKImFVVNAVYAMAHALHaMH
mGluR4a   GsksapVlrIEevAeGAVTIlpkrmsVrqFDrYFsSrtLDNNrRNiWFaEFWEDnFhCkLshalkkgshikkCTnrEriggDSaYEQEgKVqFVIdAVYAMgHALHaMH
mGluR5a   adrydVtdGyqreAvGgITIKLqSpdVkwFDdYlKlrpEtNiRNPWFqEFWqhrFqCrLegfaqenskynkCtnssl1..tthhvQDSKmgFVINAIYsMAyglHnMq
mGluR6    GskisplInIEeeAvGAITIlpkrasIdgFDqYFmtrsLENNrRNiWFaEFWEEnFnCkLtssgqsddstrkCTgeEriggDSaYEQEgKVqFVIdAVYAIAHALHsMH
mGluR7    GskinplhqEdiAeGAITIgpkratVegFDaYFSrtLENNrRNvWFaEYWEEnFnCkLtsgskkedtdrkCTgeIgrighkDSnYEQEgKVqFVIdAVYAMAHALHhMn

                    441                                                                                        550
mGluR1a   haLCPGhvGLCDAMKPIDGRKLL.DFlikssFvG........vsGeeVwFDEkGDAPGRYDImnlQyt..eanrYdYvhVGtWhEgvLnidyrkIq...mnksgmvrSVCSE
mGluR2    raLCPnthLCDAMrPVnGRiyKDFVlnvkFdapfrpadtde.VrFDtfGDgicGRYnlFtYl..ragsgrYrYqkVGyWaEg.LtLDtsfipWaspsagplPasrCSE
mGluR3    rtLCPnttkLCDAMkilDGkKLyKEYllkINFtapfnpkgAdsiVKFDtfGDgmGRYnvN7l..qtggkYslkVGhWaEt.lsLDvDslhW..srnsVrtSqcSD
mGluR4a   rdLCPGrvGLCprMdPVDGtqLLK.YIrnVNfsG......iAGnpVtFnENGDAPGRYDIYqYlrn..gsaeYkvIGsWtDh.LhLrIErmqW.pgsgqiPrSICS1
mGluR5a   msLCPGrvGLCDAMKPIDGRKLL.DslmkNfYLL......vsGdmIlFDENGDsPGRYDIlnYFkem..gkdyFdYinVGsWdngeLkMDdDeVw..skknnirVSVCSE
mGluR6    qaLCPGhtGLCpAMePtDGRtLLh.YIraVrFnG......sAGtpVmFhENGDAPGRYDIFqYaqnsasggYqaVGqWaEa.lrLmEvlrw.sgdpheVPpsqcS1
mGluR7    kdLCadyrGvCpeMeqaqGkKLLK.YIrhVNfnG......sAGtpVmFhkNGDAPGRYDIFqYQtth..ttnpgYrIlGqWtDe.LqlniEdmqW.gkqvreIPsSVCtl
```

```
mGluR1a   PCIkGqiKvirKGEvsCCWiCtaCkenEFvqDEFTCraCdlGwPnaeLTGCepIPVrYLeWsDiesIialaFsCLGILvTLFVtlIFVlIyrDTPvVKsSsRELCYIiLa
mGluR2    PClqnEvKvsvqpGEV.CCWlCiPCapYEYrlDEFTCadCglGyWPnasLTGCfeIPqeYirWgDaWAVgPVtiACLGaLATLFVlgVVRhNaTpVVKASGRELCYILLg
mGluR3    PCaPnEmKnmqpGDV.CCWiCiPCepYEYlvDEFTCmdCgpGqWPTadLsGCynIPedYiKweDaWAIgPVtiACLGfLCTciVtVFIkhNnTPIVRASGRELCYILLf
mGluR4a   PCqPGErKktvKGma..CCWhCePCtqYqvDrYTCktCpydmRPTenrTsCqpIPIvKLeWdsPWAVIfLAvvGiaATLFVvvtFVRYNDTPIVRASGRELsYVLLa
mGluR5a   PCekGqiKvirKGEvsCCWCtcPCkenEVfDEvfTCYCkaCqlGsWPTddLTGCllIPVqYLrWgDPepIaaVFACLGiLATLFVvIFIiYrDTPvVKsSsRELCYIiLa
mGluR6    PCgPGErKkmvKGvp.CCWhceaCdgYrFgvDEFTCeaCpgdmRPTpnhTGCrpPVvrLtWsSPWAalPIlLAvLGIMATtImatFmRhNDTPIVRASGRELsYVLLt
mGluR7    PCkPGqrKKtqGtp.CCWtcePCdgYqfqfDEmTCqhCpydqRPnenrTGCqnIPIiKLeWhsPWAViPYfMAmLGiATiFVmatFIRnDTPIVrASGWELsYVLLt
                                                                                                       IL I      TM I
```

```
mGluR1a   GIFLgYvcpFtLIAKpttsCyLqRLLvGLssamcYSALvTKTNrIARILaGsKKkictrkPRFmSawaQViIasiLLSVQLtIVvtliImEPPmpIlsY.......Psi
mGluR2    GVFLCYcmTFvFIAKPstaVCtLRRLgLGtafsVCYSALLTKTNRIARIFgGare..gaqrPRFISPaSOvaIcLaLLisgQLiiVaaWLvvEaPgt......gkeTaperr
mGluR3    GVsLsYcmTFFFIAKPspvICaLRRLgLGtsfaIcYSALLTKTNcIARIFDGvKn..gaqrPkFIsPsSQvfICLgLlIVQivmSvwLLiEtPgt.......ryTipekr
mGluR4a   GIFLCYatTFLMIAePdlgtCsLRRiFLGLgmsIsYaALLTKTNRiyRIFEgqKr..svsaPRFISPaSOLaIfFiLiSlQLlgIcvWFVvDPshsVvDPqdqrTIdPrf
mGluR5a   GICLgY1cTFcLIAKPkqiyCyLqRigiGLspamSYSALvTKTNRIARILaGsKKkictkkPRFmSacaQLvIaFiLLCiOLgiIvaIFImEPPdimhDY.......Psi
mGluR6    GIFLiYaiTFLMVAePcaaICaaRRLLGLgtiSYSALLTKTNRIyRIFEagKri.svtpPFISPtSQLviEFgLtSlQvvgViaWLgaqPPhsViDYeeqrTvdPeq
mGluR7    GIFLcYiiTFLMIAkPdvaVCsFRRvFLGLgmcISYaALLTKTNRIyRIFEgqKk..svtaPRLISPtSQLaItssLLISVOLiqVfiWEqvDPPniiiDYdehkTmnPeq
                               TM II                        TM III                       IL II              TM IV
```

```
mGluR1a   keVyLiCNtsnLgvVapvGyngLLImsCTyYAFKTRnvPaNFNEAKYlaFTMYTCIIWLAFvPIYFGsn.........ykiiTtCfaVSLSvtVaLGCMFtPKmYIIiakP
mGluR2    evVLrCNhrDaSMigsLAYnvLLiaICTIYAFKTRkcPENFNEAKFIGFTMYTTCIIWLAFiPIFYvTs...sdyrVQTTTMCISVSLSgVVLGCLFaPKIhILFqP
mGluR3    etVlLKCNvkDsSMisLtYdvvLVlICTvYAFKTRkcPENFNEAKFIGFTMYTCIIWLAFiPIFYvTs...sdyrVQTTTMCISVSLSgivVLGCLFaPKVhILFqp
mGluR4a   argVLKCdisDLSLIclLGYsmnLLmVtCTYYAIKTRgVPEtFNEAKpIGFTMYTCIVWLAFiPIFFGTsqsadkIyIQTTTLtVSVSLSASvsLGmLYmPKVYIILFhP
mGluR5a   reVyLiCNttnLgvVtpLGYngLLIisCTTyAFKTRnvPaNFNEAKYIaFTMYTTCIIWLAFvPIYFGsn.........ykiiTMCfsVSLSAtVaLGCMFvPKYIIiakP
mGluR6    argVLKCdmsDLSLIgcLGYsilLmVtCTTLtVSlSASyIgQTTTLtVSlSLSASvsLGmLYvPKtYVILFhP
mGluR7    argVLKCdmrDLqiicLGYsilLmVCTTYAiKTRgVPENFNEAKpIGFTMYTCIVWLAFiPIFFGTaqsaekIyIQTTTLLismLSASvaLGmLYmPKYIIiEhP
                                      TM V            IL III        TM VI                                           TM VII
```

```
mGluR1a   ErNVrsafttsdvvrmhVqdgk...ipcrsntflnifrrKkpgagnanSngkSvswsepqgrqapkgqhvwqrlsvhvktnetacnqtavikpltksyqgsgksltsda
mGluR2    qkNVv...shraptsrfgsaAprasanlqgSgsqfvptvcngrEvvDsttssl......................
mGluR3    qkNVv...thrihLnrfsvsg..tattysqsSastyvptvcngrEvlDSttssl......................
mGluR4a   EqNVp...krkrsLkavVtaAtmsnkftqknfrpngeaKselcEnlEtpalatkqtyvttythai......................
mGluR5a   ErNVrsafttstvvrmhVqdqkssssaasrsSlvnlwkrrgssgEtlsSngksvtwaqne.kstrgqhlwqrlsvhinkkenp.nqtavikpfpkstenrgpqaaaggg
mGluR6    EqNVq...krkrsLkktstmAa.....ppqnenaedaK*......................
mGluR7    ElNVq...krkrsFkavVtaAtmssrlishkpSdrpngeaKtelcEnvDpnspaakKkyvvsynnlvi......................
```

Fig. 2 *(continued on p. 8)*.

```
        991                                                                                                   1100
mGluR1a  stktlynveeedntpsahfsppspsmvvhrrgppvattpplpphltaeetplf..lads.vipkglpplpqqpqpqppqpqppqpkslmdqlqgvvtnfgsglpdfh
mGluR2   ...........................................................................................................
mGluR3   ...........................................................................................................
mGluR4a  ...........................................................................................................
mGluR5a  sgpgvagagnagctatggpeppdagpkalydvaeeeesfpaaarprspspistlshlagsagrtdddapslhsetaarsssssqg.....slmeqlssvvtftaniseln
mGluR6   ...........................................................................................................
mGluR7   ...........................................................................................................

        1101                                                                                                  1210
mGluR1a  avl....agpgtpgnslrslyppppppqhlqmlplhlstfqeesisppgediddserfkllqefvveregnteedeleeeedlptaskltpedspaltppspfrdsvas
mGluR2   ...........................................................................................................
mGluR3   ...........................................................................................................
mGluR4a  ...........................................................................................................
mGluR5a  smmlstaatpgppgtpicssylipkeiq....lpttmttfaeiqplpaievtgqaqg..........atgvspaqetptgaesapgkpdl..eelvaltppspfrdsvds
mGluR6   ...........................................................................................................
mGluR7   ...........................................................................................................

        1211                    1247
mGluR1a  gssvpsspvsesvlctppnvtyasvilrdykqsssl
mGluR2   ....................................
mGluR3   ....................................
mGluR4a  ....................................
mGluR5a  gsttpnspvsesalclpsspkydtliirdytqsssl
mGluR6   ....................................
mGluR7   ....................................
```

Fig. 2. Alignment of metabotropic glutamate receptor amino acid sequences using the GCG Pileup Program under standard conditions (Genetics Computer Group, 1991). Capital letters indicate amino acid residues that are found in at least four of the mGluR sequences at the same point of the alignment. The proposed positions of transmembrane regions (TM) (underlined) and intracellular loops (IL) are indicated. Reference: Genetics Computer Group (1991) Program Manual for the GCG Package, Version 7, April 1991, 575 Science Drive, Madison, WI 53711.

These variants are generated in two different ways: In the case of the mGluR1b, an 85-bp insertion containing an in-frame stop codon is inserted, thus leaving the remaining part of the mRNA untranslated (Pickering et al., 1993), whereas in mGluR1c, the 3' end of the mRNA is replaced by an entirely different nucleotide sequence (Pin et al., 1992). In the mGluR4b cDNA, the base pair numbers 3398–4017 have been deleted, thus leading to a different and longer C-terminal part of the protein (Simoncini et al., 1993). The alternatively spliced mGluR5b contains an insertion coding for 32 additional amino acids in the C-terminal of the protein (Minakami et al., 1993).

3. Role of the Extracellular Domain of the mGluR1a Subtype for Agonist Binding and Selectivity

3.1. Molecular Modeling

The amino-terminal region of the mGluR family is significantly larger then the corresponding portions of other G-protein-coupled receptors (Houamed et al., 1991; Masu et al., 1991; Nakanishi, 1992). This region can be subdivided into two distinct regions: an amino-terminal domain (ATD) extending for 460–480 amino acid residues, and a cysteine-rich region between the ATD and the beginning of the first membrane-spanning region. Using sequence comparison analysis and molecular modeling, O'Hara et al. (1993) investigated the ATD, and suggested that the structure of this domain and the mechanism of glutamate binding to the mGluR is similar to that reported for bacterial periplasmic binding proteins. Bacterial periplasmic binding proteins (PBPs) are receptors for high-affinity active transport of a number of amino acids, peptides, sugars, ions, and other nutrients (for a review, *see* Tam and Saier, 1993). The original observation of sequence similarity was made between the leucine/isoleucine/valine binding protein of *Pseudomonas aeruginosa* (LIVBP) (Hoshino and Kose, 1989) and the ATD of mGluR1a (O'Hara et al., 1993). Covering a distance of 372 amino acid residues, the ~20% identity observed is within the realm (Doolittle, 1987) of questionable biological significance. However, in a multiple alignment generated between mGluR1-5 and 5 members of the LIVBP family, aligned positions in which at least one amino acid residue in the LIVBP family is identical, or similar, to residue(s) in the mGluR comprise 56% of the LIVBP positions.

X-ray crystallographic structures for eight PBPs are available (for a review, *see* Quiocho, 1990), and although most are no more than 10–20% similar at the amino acid sequence level, they have been shown to fold similarly. The availability of the structural data allows for the use of sensitive sequence-structure compatibility techniques to determine relationships between sequences (Bowie et al., 1991). A model for the mGluR ligand binding site was created by a multiple-structure modeling approach involving simultaneous, but selective use of LIVBP, leucine binding protein, arabinose binding protein, and galactose binding protein coordinates (Mowbray and Petsko, 1983; Quiocho and Vyas, 1984; Sack et al., 1989a).

The model has an ellipsoid shape with two globular lobes consisting of β sheets flanked by α helices. The two lobes are separated by a cleft that contains the putative substrate binding site. The inter-lobe segments provide a base for the cleft that acts as a "hinge" between the lobes, which close on binding substrate, sequestering the substrate within the structure. When comparing the ATD model with PBPs, hydrophobic cores are preserved at the >90% level. Residues that align with binding pocket residues are 81% identical or similar; residues whose side chain is involved in ligand backbone binding are always identical. Size differences among the PBPs are accounted for by extension of variably sized loops, and almost all of the additional residues found in mGluR ATD when compared with the PBPs can be assigned to similar regions. The model allows for eight of the ten cysteines to be disulfide linked, and the remaining two probably interact with two of the remaining ten extracellular cysteines in the unmodeled domains (O'Hara et al., 1993).

Two ligand-bound forms of PBPs have been observed: an open form with ligand initially bound to one domain, and a closed form in which ligand is bound to both domains and sequestered within the cleft. The open form of LIVBP bound to leucine was studied by Sack et al. (1989a). The side chains of Ser (79) and Thr (102) were involved in hydrogen bonding to bound leucine α-ammonium and α-carboxylate atoms, and Arg(116), Phe (276), and Asp (323) were involved in LIVBP binding pocket structure. These residues align with identical residues in mGluR1-5. Ala (100) also hydrogen bonds ligand, but the bonding involves the backbone peptide carbonyl. The peptide backbone of Cys (78) is involved in van der Waals contact with the ligand. The residues that align with Cys (78), Ser (79), Ala (100), and Thr (102) in mGluR1 are Ser (164), Ser (165), Ser (186), and Thr (188).

To test their model of the ligand binding site, O'Hara et al. (1993) mutated the above residues (Ser 164, Ser 165, Ser 186, and Thr 188) to alanines, singly or in combination, and the mutant receptors were stably transfected into baby hamster kidney (BHK) cell lines and assayed for ligand affinity and phosphoinositide hydrolysis. Ala (164) and Ala (186) had little or no effect on glutamate binding or on the potencies of glutamate and quisqualate for stimulating phosphoinositide hydrolysis. Ala (165) showed a 160-fold lower potency for glutamate and an 80-fold lower potency for quisqualate for stimulating phosphoinositide hydrolysis. Although a potency for quisqualate was still discernible in Ala (188), no stimulation of phosphoinositide hydrolysis by glutamate (up to 100 m*M*) was observed with this mutant. The apparent potency for both glutamate and quisqualate was reduced at least 10,000-fold in Ala (165 + 188). The results were consistent with ligand binding modeled on LIVBP. The mGluR ATD is a member of a family of structural domains linked to a variety of receptor types, including the sea urchin guanylyl cyclase receptor, a homolog of atrial natriuretic peptide receptors (Schulz et al., 1988; Singh et al., 1988), and the ionotropic glutamate receptors (O'Hara, 1994; O'Hara et al., 1993).

3.2. Chimeric Receptors

Using chimeric receptors, in which the extracellular domain of mGluR1a is almost entirely replaced with the corresponding region of mGluR2, the structural determinants for the differing agonist selectivities of mGluR1a (quisqualate > glutamate > 1S,3R-ACPD) and mGluR2 (1S,3R-ACPD > quiqualate) have been examined by Takahashi et al., (1993). Replacement of the amino-terminal portion

(118–356) of mGluR1a with the corresponding portion of mGluR2 showed the reactivity with 1S,3R-ACPD, but not quisqualate, similar to the native mGluR2. In contrast, replacement of the amino-terminal portion (521–522) of mGluR1a with the corresponding region of mGluR2 in the reverse direction, in which a large amino-terminal portion was retained with the mGluR1a sequence, showed that this chimeric receptor reacted with quisqualate more efficiently than with glutamate, but lost the ability to interact with 1S,3R-ACPD, similar to the native mGluR1a. Thus, the incorporation of at least one-half of the amino-terminal portion of mGluR2 generates the agonist selectivity characteristic of the mGluR2 receptor. The mGluR1a receptor appears to differ from other members of the G-protein-coupled receptor family where both the extracellular domain and the transmembrane domain are involved in ligand binding and affinity (Frielle et al., 1988; Kobilka et al., 1988; Kubo et al., 1988).

mGluR1 splice variants with short C-terminal intracellular domains (mGluR1b and mGluR1c) are still able to couple to phospholipase C. This together with the previously mentioned variability in the second, but not in the third, intracellular loop from the transmembrane region formed the basis for construction of chimeric receptors between mGluR3 and mGluR1 to investigate the regions important for G-protein coupling. These experiments demonstrate that the C-terminal two-thirds of the third intracellular loop together with a region immediately following the seventh transmembrane region are necessary for the coupling to phospholipase C (Pin et al., 1993).

4. Expression of the mGluR Family in the Rodent Brain

The direct autoradiographic visualization of individual members of the mGluR family is not yet possible because of the lack of suitable ligands. Quisqualate-sensitive [^{3}H]glutamate binding sites identified in the presence of NMDA and AMPA most likely represent a summation of multiple-mGluR receptor subtypes (Cha et al., 1990; Catania et al., 1993).

Using cDNA clones encoding different mGluR receptor subtypes, the localization of receptor mRNA has been performed using *in situ* hybridization. The distribution of mRNAs encoding mGluR subtypes shows a heterogenous distribution in the central nervous system (*see* Chapter 4 for detailed description).

The cells of most brain regions show hybridization with probes for multiple mGluR subtypes. For example, the cerebral neocortex contains mRNA for mGluR1–mGluR5 and mGluR7 (Martin et al., 1992; Shigemoto et al., 1992; Kristensen et al., 1993; Nakajima et al., 1993; Ohishi et al., 1993a,b; Tanabe et al., 1993; Okamoto et al., 1994). Likewise, the pyramidal cells of the hippocampus contain mRNA for mGluR1, mGluR2, mGluR3, mGluR5, and mGluR7, but with a different intensity in the CA1–CA3 regions (Martin et al., 1992; Shigemoto et al., 1992; Nakajima et al., 1993; Ohishi et al., 1993a,b; Okamoto et al., 1994). However, only a very low level of hybridization for mGluR4 mRNA is found in the hippocampus (except for the CA2 region) (Kristensen et al., 1993; Tanabe et al., 1993). Likewise, the ventral posterior thalamic nucleus shows hybridization for mGluR1, mGluR3, mGluR4, and mGluR5 mRNA, but only very low levels of mGluR2 mRNA are found. The only mGluR subtype that shows a definitive signal in white matter tracts, such as corpus callosum, is mGluR3 (Ohishi et al., 1993b). Thus, a complicated distribution pattern is evolving for these receptors, allowing different responses to be generated in individual neurons by different patterns of mGluR subtype expression.

4.1. Changes in Expression Pattern of the mGluR Family During Development and Following Ischemia

The ability of excitatory amino acids to stimulate phosphoinositide hydrolysis in brain slices from rats in the neonatal stages is high, but declines with age (Nicoletti et al., 1986b; *see* Chapter 10). The analysis of mGluR1a mRNA, however, does not follow this pattern, but remains unchanged or increases linearly during the postnatal period depending on the region analyzed (Condorelli et al., 1992; Minakami et al., 1992).

Transient global ischemia leads to changes in the expression patterns of the mGluR subtype mRNAs: 24 h after transient global ischemia, a time-point when no neuronal damage can be observed histologically, the mRNA levels for mGluR1, mGluR2, and mGluR5 were decreased in the pyramidal cells of the CA1 and CA3 areas of the hippocampus. In contrast, mGluR4 mRNA levels were increased in these neurons, indicating that neurons may regulate pre- and post-synaptic metabotropic receptors differently in response to high extra-cellular glutamate concentrations, which are present following an ischemic insult (Meldrum and Garthwaite, 1990; Iversen et al., 1994).

5. Pharmacological Characterization of the mGluR Family

5.1. Discovery of the mGluR

The first mGluR was initially described as a G-protein-coupled receptor that on stimulation by glutamate or quisqualate activated phospholipase C, resulting in an increase in phospho-inositide hydrolysis in striatal neurons (Sladeczek et al., 1985) or in hippocampal slices (Nicoletti et al., 1986a). When *Xenopus* oocytes are injected with mRNA from whole rat brain, glutamate or quisqualate induces oscillatory calcium-activated chloride currents that are characteristics of a receptor coupled to phospholipase C (Sugijama et al., 1989).

5.2. Phosphoinositide Hydrolysis, Mobilization of Intracellular Calcium and the mGluR Family

Glutamate has been shown to stimulate calcium-activated chloride currents in *Xenopus* oocytes injected with the mRNA for mGluR1a. In mammalian cell lines expressing mGluR1a (Aramori and Nakanishi, 1992; Honoré et al., 1992; Pickering et al., 1993; Thomsen et al., 1993), agonists at mGluR1a (e.g., glutamate, quisqualate, 1S,3R-ACPD) stimulate phosphoinositide hydrolysis. Alternative splice variants of the mGluR1a (mGluR1b and mGluR1c) are also associated with alterations of the phosphoinositide/calcium-signal transduction pathway (Pin et al., 1992; Pickering et al., 1993).

In addition, mGluR5a is coupled to the phosphoinositide/calcium-signal transduction pathway when expressed in Chinese hamster ovary (CHO) cells (Abe et al., 1992). The additional members of the mGluR family (mGluR2, mGluR3, mGluR4, mGluR6, and mGluR7) are not coupled to phosphoinositide hydrolysis when expressed in mammalian cell lines (Tanabe et al., 1992, 1993; Nakajima et al., 1993; Okamoto et al., 1994).

5.3. Coupling of mGluR Family to the cAMP Cascade

In CHO cells stably expressing the mGluR1a subtype, mGluR agonists stimulate cAMP formation (Aramori and Nakanishi, 1992) with a rank order of potency similar to that seen for the stimulation of phosphoinositide hydrolysis. However, it is unknown whether mGluR1a is directly coupled to adenylyl cyclase (possibly via G_s) or whether the stimulatory effects on cAMP accumulation are the consequences of "crosstalk" between second messengers, eventually via calcium leading to the activation of Ca^{2+}/calmodulin-sensitive adenylyl cyclases or calcium/calmodulin-sensitive phosphodiesterases, as has been shown for the muscarinic receptor (Ashkenazi et al., 1989; Bonner, 1989; Felder et al., 1989).

An inhibitory coupling of mGluR2 to adenylyl cyclase was first demonstrated in CHO cells stably expressing mGluR2 (Tanabe et al., 1992). A similar inhibitory coupling of mGluR3 (Tanabe et al., 1993), mGluR4a (Thomsen et al., 1992; Tanabe et al., 1993), mGluR6 (Nakajima et al., 1993), and mGluR7 (Okamoto et al., 1994) to adenylyl cyclase has also been reported.

5.4. Arachidonic Acid Release and the mGluR Family

In CHO cells stably expressing mGluR1a, agonists at the mGluR1a subtype significantly increase [^{3}H]arachidonic acid release with a similar rank order of potency to that seen with phosphoinositide hydrolysis or cAMP formation (Aramori and Nakanishi, 1992). A coupling of mGluR5a to arachidonic acid release is not observed when the receptor is expressed in CHO cells (Abe et al., 1992). However, as with the lack of effect of mGluR5a on cAMP accumulation, it is not clear whether this reflects the diverse structural characteristics of mGluR1a and mGluR5a or differences in the actual levels of expression of receptors in the cell lines studied.

5.5. Receptor–Effector Coupling of the mGluR Family Through Multiple G-Proteins

G-Protein-coupled receptors can be coupled to their respective second-messenger systems via a variety of G-proteins (Hille, 1992). Based on the differential sensitivities of G-proteins to inactivation by the islet-activating protein of *Bordetella pertussis* (pertussis toxin), at least two types of G-proteins (termed G_p) have been implicated in the activation of phospholipase C, whereas the inhibitory actions on adenylyl cyclase are mediated by forms of G_i that are sensitive to pertussis toxin (Gilman, 1989).

The ability of glutamate to inhibit forskolin-stimulated cAMP formation via the mGluR2 (Tanabe et al., 1992), mGluR3 (Tanabe et al., 1993), mGluR4 (Kristensen et al., 1993; Tanabe et al., 1993), or mGluR6 (Nakajima et al., 1993) is completely blocked by pretreatment with low concentrations of pertussis toxin, indicating that G_i mediates their coupling to adenylyl cyclase. In contrast, the mGluR1a-mediated responses in *Xenopus* oocytes (Houamed et al., 1991; Masu et al., 1991) or in mammalian cell lines (Aramori and Nakanishi, 1992; Pickering et al., 1993; Thomsen et al., 1993a) are only partially blocked by pertussis toxin pretreatment. In similar experiments, using CHO cells expressing mGluR5a, only a small fraction of the total phosphoinositide response is sensitive to pertussis toxin (Abe et al., 1992). Thus, individual members of the mGluR family are either coupled to phospholipase C via G-proteins that are only partially sensitive to pertussis toxin or via multiple G-proteins being sensitive and insensitive to this toxin, respectively. Interestingly, an alternative spliced variant of mGluR1, mGluR1b, elicited only the pertussis toxin-insensitive component of phosphoinositide hydrolysis (Pickering et al., 1993), when expressed in BHK cells. A truncated version of mGluR1, lacking a part of the carboxy-terminal domain, was coupled to phosphoinositide hydrolysis via a pertussis toxin-insensitive G-protein when expressed in a mammalian cell line (Gabellini et al., 1993). Thus, the carboxy-terminal domain of mGluR1a may be involved in the coupling of the receptor to multiple G-proteins. On the other hand, a third version of mGluR1, mGluR1c, was partially sensitive to pertussis toxin treatment when measuring Ca^{2+}-activated chloride currents in *Xenopus* oocytes transfected with mGluR1c. Such differences in the nature of the G-protein coupling may be explained by the differences in amino acid sequences between mGluR1b and mGluR1c or, alternatively, may be related to either the different expression systems utilized or the different time periods of exposure to pertussis toxin. However, a comparison of the time-courses of the Ca^{2+}-signals following stimulation with glutamate reveals that such responses in the alternative splice variants are delayed in comparison to those of the mGluR1a type, suggesting different modes of activation of phospholipase C presumably via different types of G-proteins (Pin et al., 1992; Pickering et al., 1993).

5.6. The Regulatory Role of Protein Kinase C on the Function of the mGluR Family

Excitatory amino acid-stimulated phosphoinositide hydrolysis in hippocampal slices (Schoepp and Johnson, 1988) or in cultured cerebellar granule cells (Canonico et al., 1988) is inhibited by an activator of protein kinase C (e.g., phorbol-12,13-dibutyrate). Phorbol-12,13-dibutyrate attenuates agonist-stimulated phospho-inositide hydrolysis in mammalian cell lines stably expressing mGluR1a (Aramori and Nakanishi, 1992; Thomsen et al., 1993), suggesting a regulatory role for protein kinase C in the function of mGluR1a. Interestingly, the release of arachidonic acid in CHO cells expressing mGluR1a is potentiated by activation of protein kinase C (Aramori and Nakanishi, 1992), suggesting that these signal trans-duction pathways may be differentially regulated by protein kinase C. Whereas the molecular mechanisms for the stimulatory effects of protein kinase C activation on arachidonic acid release are unknown, similar observations have been made with subtypes of the muscarinic receptor (m1 and m3) stably expressed in A9 L cells (Conklin et al., 1988). A regulatory role of protein kinase C on the function of mGluR1a is further suggested by the observation that an inhibitor of protein kinase C, staurosporine, fully reverses the inhibitory effects of phorbol-12,13-dibutyrate at the same dose that potentiates the maximal stimulation of phosphoinositide hydrolysis in mGluR1a expressing BHK cells (Thomsen et al., 1993). In cultured cerebellar granule cells, the levels of expression of mRNA for mGluR1a were significantly diminished following prolonged exposure to quisqualate, and this effect was reversed by an activator of protein kinase C (Bessho et al., 1993). Thus, protein kinase C may regulate both the function and the expression of mGluR1a.

6. Pharmacological Characterization of the mGluR Subtypes

6.1. Pharmacological Characterization of mGluR1(a, b, c) and mGluR5a

6.1.1. Agonist Selectivities

In clonal mammalian cell lines expressing mGluR1a, mGluR agonists stimulate phosphoinositide hydrolysis with a rank order of potency of: quisqualate > glutamate > ibotenate > 1S,3R-ACPD (Aramori and Nakanishi, 1992; Pickering et al., 1993; Thomsen et al., 1993). This is in agreement with changes in calcium-activated chloride currents in *Xenopus* oocytes expressing the mGluR1a (Houamed et al., 1991; Masu et al., 1991). A similar rank order of potency was found when measuring stimulation of phosphoinositide hydrolysis in cloned cell lines expressing mGluR5a or mGluR1b (Abe et al., 1992; Pickering et al., 1993) or when measuring Ca^{2+}-dependent chloride currents in *Xenopus* oocytes injected with mRNA for mGluR1c (Pin et al., 1992). The half-maximal effective doses (EC_{50}) for stimulating phosphoinositide hydrolysis at mGluRs are shown in Table 2. Notably, the most selective mGluR agonist, 1S,3R-ACPD (Palmer et al., 1989), is only a weak partial agonist at mGluR1a, whereas a full agonist response to 1S,3R-ACPD is observed at the mGluR5a (Abe et al., 1992).

When the pharmacological profiles of mGluR1a and mGluR5a are compared with the stimulation of phosphoinositide hydrolysis in cultured neurons, cultured astrocytes, or slices of rat brain (for review, *see* Schoepp et al., 1990; Récasens et al., 1991), the high efficacy of quisqualate and a lack of sensitivity to the mGluR antagonist, L-2-amino-3-phosphonopropionate (L-AP3), correlate well. However, an additional mGluR subtype in cultured neurons, cultured astrocytes, or rat brain slices that are highly sensitive to ibotenate and L-AP3 cannot be explained by mGluR1a or mGluR5a, and suggests the presence of additional subtypes of the mGluR family coupled to phosphoinositide hydrolysis. Alternatively, the efficacy of excitatory amino acids in stimulating phosphoinositide hydrolysis in cultured neurons, cultured astrocytes, or rat brain slices may be differentially affected by their interaction with ionotropic glutamate receptors (Lonart et al., 1993).

6.1.2. Antagonists

A limited number of compounds have been shown to antagonize excitatory amino acid-evoked phosphoinositide hydrolysis in rat brain slices or in neuronal cultures, including L-AP3, L-2-amino-4-phosphonobutyrate (L-AP4), and L-serine-O-phosphate (L-SOP) (for review, *see* Schoepp et al., 1990). However, L-AP3 is a weak antagonist at mGluR1a (Aramori and Nakanishi, 1992; Pickering et al., 1993), mGluR1b (Pickering et al., 1993), and mGluR5a (Abe et al., 1992), being unable to antagonize glutamate-stimulated phosphoinositide hydrolysis fully even at millimolar concentrations. The weak antagonism of excitatory amino acid-induced phosphoinositide hydrolysis by L-AP3 in striatal neurons (Manzoni et al., 1991) may be the result of its weak agonistic effects in these neurons. However, an agonist effect of L-AP3 has not been described for any of the cloned mGluR subtypes coupled to phospholipase C (Abe et al., 1992; Aramori and Nakanishi, 1992; Manev et al., 1993; Thomsen et al., 1993). The antagonistic actions of L-AP4 are not mediated by blockade of mGluR1a (Aramori and Nakanishi, 1992; Pickering et al., 1993), mGluR1b (Pickering et al., 1993), or mGluR5a (Abe et al., 1992), but may be mediated indirectly by activation of a proposed presynaptic mGluR4 subtype (Thomsen et al., 1992; Kristensen et al., 1993; Tanabe et al., 1993). Similar observations have been made with L-SOP, which is a potent agonist at mGluR4a (Tanabe et al.,

1993; Thomsen and Suzdak, 1993a) and mGluR6 (Nakajima et al., 1993), but lacks antagonist activity at mGluR1a (Thomsen and Suzdak, 1993a).

Derivatives of phenylglycine have recently been shown to antagonize either the stimulation of phosphoinositide hydrolysis by 1S,3R-ACPD in rat cerebral cortical slices (Eaton et al., 1993) or 1S,3R-ACPD-evoked depolarizations in rat spinal cord neurons (Birse et al., 1993; Eaton et al., 1993). In rat cerebral cortical slices, (RS)-4-carboxyphenylglycine and (RS)-α-methyl-4-carboxyphenyl-glycine competitively antagonize 1S,3R-ACPD-induced phospho-inositide hydrolysis with affinities (K_B) of 1700 and 1200 μM, respectively (Eaton et al., 1993). In contrast, (S)-4-carboxy-3-hydroxyphenylglycine was a weak agonist in similar experiments (EC_{50} = 345 μM) (Birse et al., 1993). Surprisingly, (RS)-4-carboxy-3-hydroxyphenylglycine is a relatively potent competitive antago-nist at glutamate-stimulated phosphoinositide hydrolysis in BHK cells expressing mGluR1a (K_B = 29 μM) (Thomsen and Suzdak, 1993b), whereas (S)-4-carboxyphenylglycine and (RS)-α-methyl-4-carboxyphenylglycine are moderately potent competitive antago-nists at mGluR1a (K_B = 62 and 172 μM, respectively) (Thomsen et al., 1994). The competitive nature of these antagonists and the rela-tive high selectivity for mGluR1a (Birse et al., 1993; Eaton et al., 1993; Thomsen et al., 1994), suggest that these structures may prove useful for delineating the functions of mGluR1a. However, their differential functional activities at mGluR1a vs mGluR2 *(see* Section 6.2.1.), and possible effects at additional subtypes of the mGluR family, should also be considered.

6.2. Pharmacological Characterization of mGluR2 and mGluR3

6.2.1. Agonist Selectivities

Although the agonist selectivities of mGluR2 and mGluR3 are similar, striking differences are found when comparing the agonist pharmacology of mGluR2/mGluR3 with that of mGluR1a/mGluR5a. 1S,3R-ACPD is a potent agonist at mGluR2 and mGluR3, whereas quisqualate is only weakly efficacious *(see* Table 3). A rigid analog

Table 3
Potencies and Selectivities of mGluR Agonists for Subtypes of the mGluR Family[a]

	EC$_{50}$, μM						
	Glutamate	Quisqualate	Ibotenate	1S,3R-ACPD	L-AP4	L-CCG-I	References
mGluR1a	9–12	0.2–3	6–62	36–380	>1000	50	Houamed et al., 1991; Masu et al., 1991; Aramori and Nakanishi, 1992; Hayashi et al., 1992; Thomsen et al., 1993
mGluR1b	56	3	44	106	>1000	–	Pickering et al., 1993
mGluR1c	13	0.8	60	130	–	–	Pin et al., 1992
mGluR2	5	≈500	35	4	>1000	0.3	Hayashi et al. 1992; Tanabe et al., 1992
mGluR3	3	40	10	8	–	–	Tanabe et al., 1993
mGluR4a	3–5	129	590	39	0.5–1.2	50	Hayashi et al., 1992; Kristensen et al., 1993; Thomsen et al., 1992
mGluR5a	10	0.3	10	50	>1000	–	Abe et al., 1992
mGluR6	16	–	–	–	0.9	–	Nakajima et al., 1993
mGluR7	1000	–	–	–	160	–	Okamoto et al., 1994

[a]The values listed in Table 3 represent the potencies and selectivities of compounds at inducing functional responses in mGluR-transfected Xenopus oocytes or mammalian cell lines as described in the text. The EC$_{50}$ value for quisqualate at mGluR2 was estimated from a published graph (Tanabe et al. 1992). Additional agonists at mGluR1a include L-homocysteinesulphinate (Aramori and Nakanishi, 1992), 3,5-dihydroxyphenylglycine (EC$_{50}$ = 60 μM) (Ito et al., 1992) and β-N-methylamino-L-alanine (EC$_{50}$ = 480 μM) (Thomsen et al., 1993). mGluR2 is also activated by L-2(carboxycyclopropyl)-glycine-II (EC$_{50}$ = 300 μM) (Hayashi et al., 1992) and 4-carboxy-3-hydroxyphenylglycine (EC$_{50}$ = 48 μM) (Thomsen et al., 1994), whereas L-homocysteate activates mGluR4 (EC$_{50}$ = 432 μM) (Kristensen et al., 1993).

of glutamate, L-2-(carboxycyclopropyl)glycine-I (L-CCG-I) is approx 10-fold more potent than 1S,3R-ACPD at mGluR2 (Hayashi et al., 1992), leading to a rank order of potency at the mGluR2/mGluR3 of: L-CCG-I > glutamate = 1S,3R-ACPD > ibotenate > quisqualate. The relative lack of efficacy of L-CCG-I for mGluR1a and mGluR4a (Hayashi et al., 1992) suggests that this compound is a selective agonist at mGluR2 (and possibly mGluR3) and may be a useful tool for delineating the function of mGluR2 in the brain.

6.2.2. Antagonists

The antagonist pharmacologies of the mGluRs that are negatively coupled to adenylyl cyclase in the brain have not been extensively examined. In BHK cells stably expressing mGluR2, (RS)-α-methyl-4-carboxyphenylglycine and (S)-4-carboxyphenylglycine (IC_{50} = 340 and 577 μM, respectively) dose-dependently antagonized the inhibitory effects of glutamate on forskolin-stimulated cAMP formation, whereas no effect of L-AP3 was observed (Thomsen et al., 1994).

6.3. Pharmacological Characterization of mGluR4, mGluR6 and mGluR7

6.3.1. Agonist Selectivities

mGluR4, mGluR6, and mGluR7 are negatively coupled to adenylyl cyclase, thus having a signal transduction cascade in common with mGluR2 and mGluR3. However, the agonist pharmacologies of mGluR4, mGluR6, and mGluR7 are quite distinct from those of mGluR2 and mGluR3. L-AP4, and its structural analog, L-SOP, are potent agonists at mGluR4a (Thomsen et al., 1992; Tanabe et al., 1993; Thomsen and Suzdak, 1993b), mGluR6 (Nakajima et al., 1993), and mGluR7 (Okamoto et al., 1994) (see Table 3). 1S,3R-ACPD, however, is less potent at mGluR4, mGluR6, and mGluR7 (Kristensen et al., 1993; Nakajima et al., 1993; Okamoto et al., 1994; Tanabe et al., 1993). mGluR4, mGluR6, and mGluR7 are the only receptors that possess high affinities for L-AP4 and L-SOP. Based on the agonist selectivities and the regional distributions of rat brain mRNA for mGluR4a (Kristensen et al., 1993; Tanabe et al., 1993; Thomsen et al., 1992) and mGluR6 (Nakajima et al., 1993), they have been proposed as candidates for the L-AP4 receptor. The L-AP4 receptor has been postulated in order to explain the depressant action of L-AP4 on various neuronal glutamatergic pathways (Koerner and Cotman, 1981; Miller and Slaughter, 1986; Monaghan et al., 1989; Forsythe and Clements, 1990). The depressant effects of L-AP4 on synaptic transmission have been shown to involve a pertussis toxin-sensitive G-protein-coupled receptor (Thrombley and Westbrook, 1992). The effects of L-AP4 are also mimicked by the selective mGluR agonist, 1S,3R-ACPD (Baskys and Malenka, 1991; Thrombley and Westbrook, 1992; Rainnie and Shinnick-Gallagher, 1992), suggesting the involvement of the mGluR family. The mechanisms involved in a presynaptic inhibition of synaptic transmission by mGluR4a or mGluR6 have not been resolved, but may involve inhibition of Ca^{2+}-channel activities (Thrombley and Westbrook, 1992; Westbrook et al., 1993).

6.3.2. Antagonists

At present, no antagonists at the cloned subtypes mGluR4, mGluR6, and mGluR7 have been reported.

7. Conclusions and Future Research Directions

One of the most striking features of the mGluR family is its diversity in the form of multiple subtypes, second-messenger system coupling, and agonist/antagonist pharmacology. Thus far, seven subtypes of the G-protein-coupled mGluR family have been cloned, sequenced, and characterized. In addition, alternate splice variants of mGluR1, mGluR4, and mGluR5 have been identified. The multiple signal transduction pathways and heterogenous localization of the mGluR mRNA suggest that the mGluR family is involved in both the enhancement and inhibition of glutamatergic neurotransmission.

Although our knowledge of the mGluR family has greatly increased over the past several years, there are still a number of important unanswered questions:

1. Are there additional subtypes of the mGluR family? There is biochemical and electrophysiological evidence to suggest additional members of the mGluR family. Although L-AP3 has been

shown to antagonize some mGluR-mediated effects in various tissue preparations (for a review, *see* Schoepp and Conn, 1993), mGluR1–mGluR7 lack substantial sensitivity to L-AP3. In some primary tissue culture preparations (for a review, *see* Schoepp and Conn, 1993), ibotenate has been shown to stimulate phosphoinositide hydrolysis more potently than glutamate or 1S, 3R- ACPD, an effect that is not mimicked by mGluR1–mGluR7. These data suggest that additional members of the mGluR family may yet be cloned.

2. What G-proteins are involved in the coupling of subtypes of the mGluR family to signal transduction pathways in brain? Multiple G-proteins may be involved in the coupling of the mGluR family to second-messenger systems. mGluR1a-c has been shown to have differing sensitivities to pertussis toxin (Houamed et al., 1991; Masu et al., 1991; Pin et al., 1992; Pickering et al., 1993), suggesting the involvement of multiple G-proteins.

3. What are the physiological roles of alternate splice variants of the mGluR family? Recent evidence suggests that the alternate splice variants of mGluR1 (mGluR1b–c) may generate different patterns of intracellular calcium mobilization and may possess a different cellular localization (Pin et al., 1992; Pickering et al., 1993).

4. What is the central and peripheral localization of the receptor proteins of the mGluR family? Although *in situ* hybridization of mGluR1–mGluR7 has revealed a heterogenous distribution in the central nervous system, there is only very limited information available as to the regional expression of receptor protein for subtypes of the mGluR family in the central nervous system and periphery.

Finally, despite the new wealth of information obtained from mGluR1–mGluR7 expressed in *Xenopus* oocytes or mammalian cell lines, the major obstacle to a better understanding of the physiological role, and therapeutic importance of the mGluR family is the development of more potent and selective agonists and antagonists for subtypes of the mGluR family.

References

Abe, T., Sugihara, H., Nawa, H., Shigemoto, R., Mizuno, N., and Nakanishi, S. (1992) Molecular characterization of a novel metabotropic glutamate receptor mGluR5 coupled to inositol phosphate/Ca^{2+} signal transduction. *J. Biol. Chem.* **267**, 13,361–13,368.

Aramori, I. and Nakanishi, S. (1992) Signal transduction and pharmacological characteristics of a metabotropic glutamate receptor, mGluR1a, in transfected CHO cells. *Neuron* **8,** 757–765.

Ashkenazi, A., Peralta, E. G., Winslow, J. W., Ramachandran, J., and Capon, D. J. (1989) Functional diversity of muscarinic receptor subtypes in cellular signal and growth, in *Subtypes of Muscarinic Receptors IV* (Lavine, R. L. and Birdsall, N. J. M., eds.), Elsevier, New York, pp. 16–22.

Baskys, A., and Malenka, R. (1991) Agonists at metabotropic glutamate receptors presynaptically inhibit EPSCs in neonatal rat hippocampus. *J. Physiol (Lond.)* **444,** 687–701.

Baude, A., Nusser, Z., Roberts, J. D. B., Mulvihill, E., McIlhinney, R. A. J., and Somogyo, P. (1993) The 1a form of metabotropic glutamate receptor (mGluR1a) is concentrated at extra and perisynaptic membrane of discrete sub-populations of neurons as detected by immunogold reaction in the rat. *Neuron* **11,** 771–787.

Bessho, Y., Nawa, H., and Nakanishi, S. (1993) Glutamate and quisqualate regulate expression of metabotropic glutamate receptor messenger RNA in cultured cerebellar granule cells. *J. Neurochem.* **60,** 253–259.

Birse, E., Eaton, S., Jane, D., Jones, P., Porter, R., Pook, P., Sunter, D., Udvarhelyi, P., Wharton, B., Roberts, P., and Watkins, J. (1993) Phenylglycine derivatives as new pharmacological tools for investigating the role of metabotropic glutamate receptors in the central nervous system. *Neurosci.* **52,** 481–488.

Bonner, T. (1989) New subtypes of muscarinic acetylcholine receptors, in *Subtypes of Muscarinic Receptors IV* (Lavine, R. L. and Birdsall, N. J. M., eds.), Elsevier, New York, pp. 11–15.

Bowie, J. W., Luthy, R., and Eisenberg, D. (1991) A method to identify protein sequences that fold into a known three-dimensional structure. *Science* **252,** 164–170.

Canonico, P. L., Favit, A., Catania, M. V., and Nicoletti, F. (1988) Phorbol esters attenuate glutamate-stimulated inositol phospholipid hydrolysis in neuronal cultures. *J. Neurochem.* **51,** 1049–1053.

Catania, M., Hollingsworth, Z., and Young, A. (1993) Quisqualate resolves 2 distinct metabotropic [3-H] glutamate binding-sites. *Neurorep.* **4,** 311–313.

Cha, J. J., Makowiec, R. L., Penney, J. B., and Young A. B. (1990) L-[3H]glutamate labels the metabotropic excitatory amino acid receptor in rodent brain. *Neurosci. Lett.* **113,** 78–83.

Condorelli, D., Dellalbani, P., Amico, C., Casabona, G., Genazzani, A., Sortino, M., and Nicoletti, F. (1992) Developmental Profile of Metabotropic Glutamate Receptor Messenger RNA in Rat Brain. *Mol. Pharmacol.* **41,** 660–664.

Conklin, B. R., Brann, M. R., Buckley, N. J., Ma, A. L., and Bonner, T. I. (1988) Stimulation of arachidonic acid release and inhibition of mitogenesis by cloned genes for muscarinic receptor subtypes stably expressed in A9 L cells. *Proc. Natl. Acad. Sci. USA* **85,** 8698–8702.

Doolittle, R. F. (1987) *Of urfs and orfs: a primer on how to analyze devised amino acid sequences.* University Science Books, Mill Valley, CA, pp. 11,12.

Eaton, S. A., Jane, D. E., St. J. Jones, P. L., Porter, R. H. P., Pook, P. C.-K., Sunter, D. C., Udvarhelyi, P. M., Roberts, P. J., Salt, T. E., and Watkins, J. C. (1993)

Competitive antagonism at metabotropic glutamate receptors by (S)-4-carboxy-phenylglycine and (RS)-α-methyl-4-carboxyphenylglycine. *Eur. J. Pharmacol. Mol. Pharmacol. Sect.* **244**, 195–214.

Felder, C. C., Kanterman, R. Y., Ma, A. L., and Axelrod, J. (1989) A transfected m1 muscarinic acetylcholine receptor stimulates adenylate cyclase via phosphatidylinositol hydrolysis. *J. Biol. Chem.* **264**, 20,356–20,362.

Forsythe, I. D. and Clements, J. D. (1990) Presynaptic glutamate receptors depress excitatory monosynaptic transmission between mouse hippocampal neurones. *J. Physiol. (Lond.)* **429**, 1–16.

Franke, R. R., König, B., Sakmar, T. P., Khorana, H. G., and Hofmann, K. P. (1990) Rhodopsin mutants that bind but fail to activate transduction. *Science* **250**, 123–125.

Frielle, T., Daniel, K. W., Caron, M. G., and Lefkowitz, R. J. (1988) Structural basis of beta-adrenergic receptor subtype specificity studied with chimeric beta 1/beta 2 adrenergic receptors. *Proc. Natl. Acad. Sci. USA* **85**, 9494–9498.

Gabellini, N., Manev, R., Candeo, P., and Manev, H. (1993) Carboxyl domain of the glutamate receptor directs its coupling to metabolic pathways. *Neurorep.* **4**, 531–534.

Gilman, A. G. (1989) G Proteins and regulation of adenylyl cyclase. *JAMA* **262**, 1819–1825.

Hayashi, Y., Tanabe, Y., Aramori, I., Masu, M., Shimamoto, K., Ohfune, Y., and Nakanishi, S. (1992) Agonist analysis of 2-(carboxycyclopropyl)glycine isomers for cloned metabotropic glutamate receptor subtypes expressed in Chinese hamster ovary cells. *Br. J. Pharmacol.* **107**, 539–543.

Headley, P. M. and Grillner, S. (1990) Excitatory amino acids and synaptic transmission: the evidence for a physiological function. *Trends Pharmacol. Sci.* **11**, 205–211.

Hille, B. (1992) G Protein-coupled mechanisms and nervous signaling. *Neuron* **9**, 187–195.

Honoré, T., Petersen, V., Suzdak, P., Thomsen, C., and Mulvihill, E. (1992) Configurations of non-NMDA glutamate receptors. *Mol. Neuropharmacol.* **2**, 61–64.

Hoshino, T. and Kose, K. (1989) Cloning and nucleotide sequence of BarC, the structural gene for the Leucine-, isoleucine- and valine-binding protein of Pseudomonas aeruginosa. *J. Bacteriol.* **171**, 6300–6306.

Houamed, K. M., Kuijper, J. L., Gilbert, T. L., Haldeman, B. A., O'Hara, P. J., Mulvihill, E. R., Almers, W., and Hagen, F. S. (1991) Cloning, expression and gene structure of a G protein-coupled glutamate receptor from rat brain. *Science* **252**, 1318–1321.

Ito, I., Kohda, A., Tanabe, S., Hirose, E., Hayashi, M., Mitsunaga, S., and Sugiyama, H. (1992) 3,5-Dihydroxyphenylglycine—A potential agonist of metabotropic glutamate receptors. *Neurorep.* **3**, 1013–1016.

Iversen, L., Mulvihill, E., Haldeman, B., Diemer, N. H., Kaiser, F., Sheardown, M. J., and Kristensen, P. (1994) Changes in metabotropic glutamate receptor mRNA levels following global ischemia: Increase of a putative presynaptic subtype (mGluR4) in highly vulnerable brain areas. *J. Neurochem.* (in press).

Kobilka, B. K., Kobilka, T. S., Daniel, K., Regan, J. W., Caron, M. G., and Lefkowitz, R. J. (1988) Chimeric alpha-1, beta-2 adrenergic receptors: delineation of domains involved in effector coupling and ligand binding specificity. *Science* **240**, 1310–1316.

Koerner, J. F. and Cotman, C. W. (1981) Micromolar L-2-amino-4-phosphonobutyric acid selectively inhibits perforant path synapses from lateral entorhinal cortex. *Brain Res.* **216,** 192–197.

Kristensen, P., Suzdak, P. D., and Thomsen, C. (1993) Expression pattern and pharmacology of the rat type IV metabotropic glutamate receptor. *Neurosci. Lett.* **155,** 159–162.

Kubo, T., Bujo, H., Nakai, I., Mishina, M., and Numa, S. (1988) Location of a region of the muscarinic acetylcholine receptor involved in selective effector coupling. *FEBS Lett.* **241,** 119–125.

Lonart, G., Alagarsamy, S., and Johnson, K. M. (1993) (R,S)-α-amino-3-hydroxy-5-methylisoxazole-4-propionic acid (AMPA) receptors mediate a calcium-dependent inhibition of the metabotropic glutamate receptor-stimulated formation of inositol 1,4,5-trisphosphate. *J. Neurochem.* **60,** 1739.

Luttrell, L. M., Ostrowski, J., Cotecchia, S., Kendall, H., and Lefkowitz, R. J. (1993) Antagonism of catecholamine receptor signaling by expression of cytoplasmic domains of the receptors. *Science* **259,** 1453–1456.

MacDermott, A. B., Mayer, M. L., Westbrook, G. L., Smith, S. L., and Barker, J. L. (1986) NMDA-receptor activation increases cytoplasmic calcium concentration in cultured spinal cord neurons. *Nature* **321,** 519–522.

Maggio, R., Vogel, Z., and Wess, J. (1993) Reconstitution of functional muscarinic receptors by co-expression of amino- and carboxyl-terminal receptor fragments. *FEBS Lett.* **319,** 195–200.

Manev, R. M., Favaron, M., Gabellini, N., Candeo, P., and Manev, H. (1993) Functional evidence for a L-AP3-sensitive metabotropic receptor different from glutamate metabotropic receptor mGluR1. *Neurosci. Lett.* **155,** 73–76.

Manzoni, O. J. J., Poulat, F., Do, E., Sahuquet, A., Sassetti, I., Bockaert, J., and Sladeczek, F. (1991) Pharmacological characterization of the quisqualate receptor coupled to phospholipase C (Q_p) in striatal neurons. *Eur. J. Pharmacol.* **207,** 231–241.

Martin, L. J., Blackstone, C. D., Huganir, R. L., and Price, D. L. (1992) Cellular localization of a metabotropic glutamate receptor in rat brain. *Neuron* **9,** 259–270.

Masu, M., Tanabe, Y., Tsuchida, K., Shigemoto, R., and Nakanishi, S. (1991) Sequence and expression of a metabotropic glutamate receptor. *Nature* **349,** 760–765.

Mayer, M. L. and R. J. Miller (1990) Excitatory amino acid receptors. Second messengers and regulation of intracellular calcium in mammalian neurons. *Trends Pharmacol. Sci.* **11,** 254–265.

Meldrum, B. and Garthwaite, J. (1990) Excitatory amino acid neurotoxicity and neurodegenerative disease. *Trends Pharmacol. Sci.* **11,** 379–388.

Miller, R. F. and Slaughter, M. M. (1986) Excitatory amino acid receptors of the retina: diversity of subtypes and conductance mechanisms. *Trends Neurosci.* **9,** 211–218.

Minakami, R., Katsuki, F., and Sugiyama, H. (1993) A variant of metabotropic glutamate receptor subtype 5: an evolutionary conserved insertion with no termination codon. *Biochem. Biophys. Res. Comm.* **194,** 622–627.

Minakami, R., Hirose, E., Yoshioka, K., Yoshimura, R., Misumi, Y., Sakaki, Y., Tohyama, M., Kiyama, H., and Sugiyama, H. (1992) Postnatal development of messenger RNA specific for a metabotropic glutamate receptor in the rat brain. *Neurosci. Res.* **15**, 58–63.

Monaghan, D. T., Bridges, R. J., and Cotman, C. W. (1989) The excitatory amino acid receptors: their classes, pharmacology, and distinct properties in the function of the central nervous system. *Annu. Rev. Pharmacol. Toxicol.* **29**, 365–402.

Mowbray, S. L. and Petsko, G. A. (1993) The X-ray structure of the periplasmic galactose binding protein from Salmonella typhimurium at 3.0 A resolution. *J. Biol. Chem.* **259**, 7991–7997.

Murphy, S. N., Thayer, S. A., and Miller, R. J. (1987) The effects of excitatory amino acids on intracellular calcium in single mouse striatal neurons in-vitro. *J. Neurosci.* **7**, 4145–4152.

Nahorski, S. R. and Potter, B. V. L. (1989) Molecular recognition of inositol polyphosphates by intracellular receptors and metabolic enzymes. *Trends Pharmacol. Sci.* **10**, 139–146.

Nakajima, Y., Iwakabe, H., Akazawa, C., Nawa, H., Shigemoto, R., Mizuno, N., and Nakanishi, S. (1993) Molecular characterization of a novel retinal metabotropic glutamate receptor mGluR6 with a high agonist selectivity for L-2-amino-4-phosphonobutyrate. *J. Biol. Chem.* **268**, 11,868–11,873.

Nakanishi, S. (1992) Molecular diversity of glutamate receptors and implications for brain function. *Nature* **258**, 597–603.

Nicoletti, F., Meek, J. L., Iadorola, M. J., Chaung, D. M., Roth, B. L., and Costa, E. (1986a) Coupling of inositol phospholipid metabolism with excitatory amino acid recognition sites in rat hippocampus. *J. Neurochem.* **50**, 1605–1613.

Nicoletti, F., Iadarola, M. J., Wroblewski, J. T., and Costa, E. (1986b) Excitatory amino acid recognition sites coupled with inositol phospholipid metabolism: developmental changes and interaction with alpha1-adrenoceptors. *Proc. Natl. Acad. Sci. USA* **83**, 1931–1935.

O'Hara, P. J. (1994) Metabotropic glutamate receptors in *Biomembranes,* vol. 2, in press.

O'Hara, P. J., Sheppard, P. O., Thøgersen, H., Venzia, D., Haldeman, B. A., McGrane, V., Houamed, K. M., Thomsen, C., Gilbert, T. L., and Mulvihill, E. R. (1993) The ligand binding domain in metabotropic glutamate receptors belongs to a family of receptor structures related to bactorial periplasmic binding proteins. *Neuron* **11**, 41–52.

Ohishi, H., Shigemoto, R., and Mizuno, N. (1993a) Distribution of the messenger-RNA for a metabotropic glutamate receptor, mGluR2, in the central-nervous-system of the rat. *Neuroscience* **53**, 1009–1018.

Ohishi, H., Shigemoto, R., and Mizuno, N. (1993b) Distribution of the mRNA for a metabotropic glutamate receptor (mGluR3) in the rat brain, an *in-situ* hybridization study. *J. Comp. Neurol.* **335**, 252–266.

Okamoto, N., Seiji, H., Akazawa, C., Hayashi, Y., Shigemoto, R., Mizuno, N., and Nakanishi, S. (1994) Molecular characterization of a new metabotropic glutamate receptor mGluR7 coupled to inhibitory cyclic AMP signal transduction. *J. Biol. Chem.* **269**, 1231–1236.

Palmer, E., Monaghan, D. T., and Cotman, C. W. (1989) Trans-ACPD, a selective agonist of the phosphoinositide-coupled excitatory amino acid receptor. *Eur. J. Pharmacol.* **166,** 585–594.

Pickering, D. S., Thomsen, C., Suzdak, P. D., Fletcher, E. J., Robitaille, R., Salter, M. W., MacDonald, J. F., Huang, X.-P., and Hampson, D. R. (1993) A comparison of two alternatively spliced forms of a metabotropic glutamate receptor coupled to phosphoinositide turnover. *J. Neurochem.* **61,** 85–92.

Pin, J. P., Joly, C., Heinemann, S. F., and Bockaert, J. (1994) Functional roles of intracellular domains of metabotropic glutamate receptors. *EMBO J.* in press.

Pin, J. P., Waeber, C., Prezeau, L., Bockaert, J., and Heinemann, S. F. (1992) Alternative splicing generates metabotropic glutamate receptors inducing different patterns of calcium release in Xenopus Oocytes. *Proc. Natl. Acad. Sci. USA* **89,** 10,331–10,335.

Quiocho, F. A. and Vyas, N. K. (1984) Novel stereospecificity of the L-arabonise-binding protein. *Nature* **310,** 381–386.

Quiocho, F. A. (1990) Atomic structures of periplasmic binding proteins and the high-affinity active transport systems in bacteria. *Phil. Trans. R. Soc. Lond. B* **326,** 341–351.

Rainnie, D. G. and Shinnick-Gallagher, P. (1992) Trans-ACPD and L-APB presynaptically inhibit excitatory glutamergic transmission in the basolateral amygdala (BLA). *Neurosci. Lett.* **139,** 87–91.

Récasens, M., Mayat, E., and Guiramand, J. (1991) Excitatory amino acid receptors and phosphoinositid breakdown: Facts and perspectives, in *Current Aspects of the Neurosciences,* vol. 3 (Osborne, N. N., ed.), Macmillan, New York, pp. 103–175.

Sack, J. S., Saper, M. A., and Quiocho, F. A. (1989a) Periplasmic binding protein structure and function. Refined X-ray structures of the leucine/isoleucine/valine-binding protein and its complex with leucine. *J. Mol. Biol.* **206,** 171–191.

Sack, J. S., Trakhanov, S. D., Tsigannik, I. H., and Quiocho, F. A. (1989b) Structure of the L-leucine-binding protein refined at 2.4 A resolution and comparison with the Leu/Ile/Val-binding protein structure. *J. Mol. Biol.* **206,** 193–207.

Schoepp, D. D. and Conn, P. J. (1993) Metabotropic glutamate receptors in brain function and pathology. *Trends Pharmacol. Sci.* **14,** 13–20.

Schoepp, D. D. and Johnson, B. G. (1988) Selective inhibition of excitatory amino acid-stimulated phosphoinositide hydrolysis in the rat hippocampus by activation of protein kinase C. *Biochem. Pharmacol.* **37,** 4299–4305.

Schoepp, D. D., Bockaert, J., and Sladeczek, F. (1990) Pharmacological and functional characteristics of metabotropic excitatory amino acid receptors. *Trends Pharmacol. Sci.* **11,** 508–515.

Schulz, S., Lowe, D. G., Thorpe, D. S., Rodriguez, H., Kuang, W. J., Dangott, L. J., Chinkers, M., Goeddel, D. V., and Garbers, D. L. (1988) Membrane guanylate cyclase is a cell-surface receptor with homology to protein kinase. *Nature* **334,** 708–712.

Shigemoto, R., Nakanishi, S., and Mizuno, N. (1992) Distribution of the messenger RNA for a metabotropic glutamate receptor (mGluR1) in the central nervous system—An *In situ* hybridization study in adult and developing rat. *J. Comp. Neurol.* **322,** 121–135.

Simoncini, L., Haldeman, B. A., Yamagiwa, T., and Mulvihill, E. R. (1993) Functional characterization of metabotropic glutamate receptor subtypes. *Biophys. J.* **64**, A84.

Singh, S., Lowe, D. G., Thorpe, D. S., Rodriguez, H., Kuang, W. J., Dangott, L. J., Chinkers, M., Goeddel, D. V., and Garbers, D. L. (1988) Membrane guanylate cyclase is a cell-surface receptor with homology to protein kinase. *Nature* **334**, 708–712.

Sladeczek, F., Pin, J.-P., Récasens, M., Bockaert, J., and Weiss, S. (1985) Glutamate stimulates inositol phosphate formation in striatal neurons. *Nature* **317**, 717–719.

Sugijama, H., Ito, I., and Watanabe, M. (1989) Glutamate receptor subtypes may be classified into two major categories: A study on *Xenopus* oocytes injected with rat brain mRNA. *Neuron* **3**, 129–132.

Takahashi, K., Tsuchida, K., Taneba, Y., Masu, M., and Nakanishi, S. (1993) Role of the large extracellular domain of metabotropic glutamate receptors in agonist selectivity determination. *J. Biol. Chem.* **268**, 19,341–19,345.

Tam, R. and Saier, M. H. (1993) Structural, functional, and evolutionary relationships among extracellular solute-binding receptors of bacteria. *Microbiol. Rev.* **57**, 320–346.

Tanabe, Y., Masu, M., Ishii, T., Shigemoto, R., and Nakanishi, S. (1992) A family of metabotropic glutamate receptors. *Neuron* **8**, 169–179.

Tanabe Y., Nomura, A., Masu, M., Shigemoto, R., and Nakanishi, S. (1993) Signal transduction, pharmacological properties, and expression patterns of two rat metabotropic glutamate receptors, mGluR3 and mGluR4. *J. Neurosci.* **13**, 1372–1378.

Thomsen, C. and Suzdak, P. D. (1993a) L-Serine-*O*-phosphate has affinity for the type IV, but not the type I, metabotropic glutamate receptor. *Neurorep.* **4**, 1099–1101.

Thomsen, C. and Suzdak, P. D. (1993b) 4-Carboxy-3-hydroxyphenylglycine, an antagonist at type I metabotropic glutamate receptors. *Eur. J. Pharmacol. Mol. Pharmacol. Sect.* **245**, 299–301.

Thomsen, C., Kristensen, P., Mulvihill, E., Haldeman, B., and Suzdak, P. D. (1992) L-2-Amino-4-phosphonobutyrate (L-AP4) is an agonist at the type IV metabotropic glutamate receptor which is negatively coupled to adenylate cyclase. *Eur. J. Pharmacol. Mol. Pharmacol. Sect.* **227**, 361–363.

Thomsen, C., Mulvihill, E. R., Haldeman, B., Pickering, D. S., Hampson, D. R., and Suzdak, P. D. (1993) A pharmacological characterization of the mGluR1a subtype of the metabotropic glutamate receptor expressed in a cloned baby hamster kidney cell line. *Brain Res.* **619**, 22–28.

Thomsen C., Boel, E., and Suzdak, P. D. (1994) Actions of phenylglycine analogs at subtypes of the metabotropic glutamate receptor family. *Eur. J. Pharmacol. Mol. Pharmacol. Sect.* **267**, 77–84.

Thrombley, P. Q. and Westbrook, G. L. (1992) L-AP4 inhibits calcium currents and synaptic transmission via a G-protein-coupled glutamate receptor. *J. Neurosci.* **12**, 2043–2050.

Westbrook, G. L., Sahara, Y., Saugstad, J. A., Kinzie, J. M., and Segerson, T. P. (1993) Regulation of ion channels by ACPD and AP4. *Funct. Neurol.* **8(4)**, 56.

CHAPTER 2

Pharmacological Properties of Metabotropic Glutamate Receptors

Darryle D. Schoepp

1. Recognition of a Glutamate Receptor Class with Novel Pharmacological Properties

1.1. Glutamate Receptors Linked to Phosphoinositide Hydrolysis

Metabotropic glutamate receptors (mGluRs) were initially discovered by their unique coupling mechanism and pharmacological characteristics. This was preceded by the recognition in the early 1980s of phosphoinositide hydrolysis as a novel signal transduction pathway in the mammalian central nervous system (CNS) (Berridge and Irvine, 1984; Nishizuka, 1984). Pharmacological studies of receptor-mediated phosphoinositide hydrolysis in CNS tissues were greatly facilitated by use of lithium ion to amplify agonist-dependent responses. Lithium at concentrations of 1–10 mM uncompetitively inhibits the enzyme inositol-1-monophosphatase (Hallcher and Sherman, 1980). Using ^{3}H-myo-inositol to label ^{3}H-phosphoinositi-

The Metabotropic Glutamate Receptors Eds.: P. J. Conn and J. Patel
© 1994 Humana Press Inc., Totowa, NJ

des and lithium to inhibit inositol-1-monophosphatase, Berridge et al. (1982) demonstrated in rat cerebral cortical slices that cholinergic and adrenergic receptor agonists will increase phosphoinositide hydrolysis and, thus, produce an easily measured increase in the formation of ^{3}H-inositol-1-monophosphate. The use of this sensitive technique allowed other investigators to begin characterizing the various receptor systems that were linked to this novel second-messenger system in the CNS (*see* Fisher and Agranoff, 1987).

Using the lithium technique to measure phosphoinositide hydrolysis, the excitatory amino acid agonists L-glutamate, quisqualate, and ibotenate were shown to activate phosphoinositide hydrolysis in a number of tissues, including striatal neurons in culture (Sladeczek et al., 1985), rat brain slices (Bardsley and Roberts, 1983; Nicoletti et al., 1986b), synaptoneurosomes from the rat brain (Recasens et al., 1987), glial cells in culture (Pearce et al., 1986), and cultured retinal cells (Osborne, 1990). Pharmacological data indicated that the effects of these agonists on phosphoinositide hydrolysis were not the result of secondary events subsequent to the activation of ion channel-linked (ionotropic) glutamate receptors, such as calcium entry activating phospholipase C. The selective ionotropic agonists N-methyl-D-aspartate (NMDA) and kainate, which effectively increase gating of extracellular calcium into the cell, only weakly increase phosphoinositide hydrolysis in these tissues (Sladeczek et al., 1985; Nicoletti et al., 1986b; Recasens et al., 1987). Furthermore, although quisqualate is the most potent agonist in activating phosphoinositide hydrolysis in many tissues, this activity was not mimicked by the compound α-amino-3-hydroxy-5-methyl-4-isoxazolepropionic acid (AMPA) (Palmer et al., 1988; Recasens et al., 1988; Schoepp and Johnson, 1988; Weiss, 1989; Manzoni et al., 1991a). Thus, it was recognized that AMPA is a more selective ionotropic "quisqualate" receptor agonist than quisqualate itself. For this reason, "quisqualate" ionotropic receptors are now termed AMPA receptors (*see* Monaghan et al., 1989).

Pharmacological studies with ionotropic glutamate receptor antagonists also supported the hypothesis of a novel "metabotropic" glutamate receptor coupled to phosphoinositide hydrolysis. For example, ibotenate is a potent NMDA receptor agonist that is highly efficacious in stimulating phosphoinositide hydrolysis in CNS tissues. Activation of phosphoinositide hydrolysis by ibotenate was not inhibited by either competitive (i.e., D-2-amino-5-phosphono-pentanoic acid, AP5) or noncompetitive (i.e., MK 801) NMDA receptor antagonists (*see* Schoepp et al., 1990a; Schoepp and Conn, 1993). Likewise, stimulation by the AMPA receptor agonist quisqualate was not antagonized by quinoxalinedione compounds, such as 6,7-dinitroquinoxaline-2,3-dione (DNQX) and 6-cyano-7-nitroquinoxaline-2,3-dione (CNQX) (Palmer et al., 1988; Recasens et al., 1988; Schoepp and Hillman, 1990; Manzoni et al., 1991a), which are potent and selective competitive antagonists against ionotropic responses produced by both AMPA and quisqualate (Honore et al., 1988; Watkins et al., 1990).

1.2. Phosphoinositide-Linked Glutamate Receptors Mobilize Intracellular Calcium Through a "Metabotropic" Mechanism

Receptor-mediated phosphoinositide hydrolysis can also be studied in *Xenopus* oocytes injected with rat brain messenger RNA. Phosphoinositide hydrolysis leads to the formation of inositol-1,4,5-trisphosphate and the mobilization of intracellular calcium, and this activates a calcium-sensitive oscillatory chloride current in this preparation through a GTP binding protein-dependent (pertussis toxin-sensitive) mechanism. The work of Sugiyama et al. (1987, 1989) showed that L-glutamate, quisqualate, and ibotenate, but not NMDA, AMPA, or kainate, produce oscillatory currents in this preparation that are inhibited by pertussis toxin. This work provided the first direct evidence that mGluRs were "metabotropic" or directly coupled via G-proteins to activation of phospholipase C. The later use of *Xenopus* oocytes in this manner also provided the technical means to clone the first mGluR protein. The mobilization of intracellular calcium subsequent to *in situ* mGluR activation has also been shown in neurons and glia using calcium imaging with fura-2 (*see* Mayer and Miller, 1990). In cultured hippocampal neurons (Murphy and Miller, 1988) and astrocytes (Ahmed et al., 1990) L-glutamate and quisqualate, but not AMPA or NMDA, produce calcium transients in the absence of extracellular calcium. It is likely that these quisqualate-sensitive AMPA-insensitive calcium responses are a consequence of *in situ* mGluR-activated phosphoinositide hydrolysis.

1.3. Selective Activation of mGluRs with 1S,3R-ACPD

Although early pharmacological studies clearly indicated the existence of a novel class of glutamate receptors linked to second-messenger formation, the use of agonists, such as quisqualate and ibotenate, to study the function of mGluRs was hampered by their concomitant ionotropic activities. However, it was subsequently discovered that the glutamate analog ($\pm$)*trans*-1-amino-1,3-dicarboxylic acid (*trans*-ACPD) (*see* Section 2.2.) stimulated phosphoinositide hydrolysis in brain slices at concentrations having no effect on ionotropic glutamate receptors (Palmer et al., 1989; Desai and Conn, 1990; Manzoni et al., 1990; Schoepp and Hillman, 1990; Schoepp et al., 1991a). Like quisqualate or ibotenate, *trans*-ACPD also evokes oscillatory currents in oocytes injected with rat brain mRNA (Watson et al., 1990; Tanabe et al., 1991), and mobilizes intracellular calcium in neurons (Irving et al., 1992) and glia (Holzwarth et al., 1993).

Trans-ACPD is a racemic mixture of 1S,3R- and 1R,3S- stereoisomers (*see* Schoepp et al., 1990a). It is the 1S,3R-ACPD isomer which is responsible for mGluR activation by the racemic *trans*-ACPD. In many systems, 1S,3R-ACPD is clearly more potent and efficacious than 1R,3S-ACPD (for reviews, *see* Schoepp et al., 1990a; Schoepp and Conn, 1993). This stereoselectivity has been found when studying the activation of phosphoinositide hydrolysis in rat brain slices (Schoepp et al., 1991b) and cultured neurons (Aronica et al., 1993), and mobilization of intracellular calcium in cultured neurons (Irving et al., 1990). 1S,3R-ACPD is generally active at micromolar concentrations, but does not activate ionotropic glutamate receptors at concentrations up to 1–3 mM (Sacaan and Schoepp, 1992; Schoepp, 1993).

This high selectivity of 1S,3R-ACPD makes it very useful as a pharmacological tool to study the cellular and behavioral consequences of *in situ* mGluR activation. Numerous 1S,3R-ACPD-induced cellular/behavior effects have been found suggesting the involvement of these receptors and, presumably, the activation of phosphoinositide hydrolysis. However, many of the effects of 1S,3R-ACPD cannot be explained by the activation of this second-messenger pathway (*see* Schoepp and Conn, 1993). For example, the reduction of potassium conductances in rat CA3 pyramidal neurons subsequent to mGluR activation with *trans*-ACPD is not affected by protein kinase C inhibition (Gerber et al., 1992). Likewise, inhibition of phosphoinositide hydrolysis by L-2-amino-3-phosphonopropionate (L-AP3) in rat hippocampal slices has no effect on the physiological responses measured in the rat CA1 region (Desai et al., 1992). Furthermore, modulation of hippocampal pyramidal cell excitability to *trans*-ACPD is not regulated developmentally in the same manner as *trans*-ACPD-induced phosphoinositide hydrolysis (Boss et al., 1992). However, it is now known that 1S,3R-ACPD alters other second-messenger pathways *in situ*, in addition to increasing phosphoinositide hydrolysis (*see* Chapter 3). These include increases in phospholipase D activity (Boss and Conn, 1992; Holler et al., 1993), decreases in cAMP formation (Cartmell et al., 1992; Casabona et al., 1992; Schoepp et al., 1992), increases in cAMP formation (Casabona et al., 1992; Winder and Conn, 1992), and increases in cGMP formation (Okada, 1992). Furthermore, from the cloning of mGluRs, it is now established that mGluRs are heterogeneous and coupled to multiple second-messenger systems. Both the activation of phosphoinositide hydrolysis and inhibition of adenylyl cyclase have been demonstrated in cells expressing cloned mGluRs (*see* Nakanishi, 1992). It is now important to sort out the molecular basis for the many reported effects of 1S,3R-ACPD, including the identification of the particular mGluR involved and its relevant second-messenger pathway.

1.4. Cloning of a Family of mGluRs Coupled to Multiple Second Messengers

The recognition that glutamate receptors coupled to phosphoinositide hydrolysis could be readily expressed and studied using the

Xenopus oocyte system provided everything necessary to clone the first mGluR. This receptor protein, now termed mGluR1a, was independently discovered by two groups using expression cloning from a rat cerebellar library (Masu et al., 1991; Houamed et al., 1991). mGluR1a possesses similar pharmacological characteristics to phosphoinositide-coupled mGluR responses in rat brain tissues. When expressed in the *Xenopus* oocyte or in nonneuronal cells or such as Chinese hamster ovary (CHO) cells (Aramori and Nakanishi, 1992) or baby hamster kidney (BHK) cells (Thomsen et al., 1993), mGluR1a is activated by quisqualate and glutamate at low-micromolar concentrations. 1S,3R-ACPD (as *trans*-ACPD) also activates mGluR1a, but requires high-micromolar to millimolar concentrations.

Armed with probes based on mGluR1a sequence, a heterogeneous family of mGluRs was rapidly cloned by homology screening (Nakanishi, 1992). These include three splice variants forms of mGluR1 (α, β, and c) that are each coupled to phosphoinositide hydrolysis when expressed. Another phosphoinositide-coupled mGluR is rat mGluR5, which is about 70% homologous to mGluR1 and is present in a and b splice variant forms (Abe et al., 1992; Minakami et al., 1993). A number of clones with lower (about 40%) homology to mGluR1 and mGluR5 have also been found, and include rat mGluR2 (Tanabe et al., 1992), mGluR3 (Tanabe et al., 1992), mGluR4 (Tanabe et al., 1992), mGluR6 (Nakajima et al., 1993), and mGluR7 (*see* Chapter 1). These latter mGluRs do not readily couple to phosphoinositide hydrolysis, as demonstrated by lack of agonist-induced currents when they are expressed in *Xenopus* oocyte. However, they are negatively coupled to adenylyl cyclase when they are expressed in nonneuronal cells, as demonstrated by agonist-induced inhibition of forskolin-stimulated cAMP formation. The negatively coupled cAMP-linked mGluRs can also be subdivided into two groups, with each group having higher sequence homology with each other and similar pharmacology. Rat mGluR2 and mGluR3 are 70% homologous, and are each potently activated by *trans*-ACPD. Rat mGluR4 and mGluR6 share higher homology and are both insensitive to *trans*-ACPD, but are potently activated by L-2-amino-4-phosphonobutyrate (L-AP4) (*see* Nakanishi, 1992; *see also* Chapter 1).

The ability to clone, express, and study the pharmacology and distribution of members of this new and novel receptor family has rapidly increased knowledge of mGluR functions. Comparison of pharmacological information using cloned mGluRs to *in situ* mGluR responses is a powerful way to increase this knowledge. The expression of cloned mGluRs also offers a means to discover highly mGluR subtype-specific agonists and antagonists, which can then be applied as *in situ* pharmacological tools and potential therapeutic agents.

2. Compounds That Act on mGluRs

2.1. Nonselective (Ionotropic/Metabotropic) Receptor Agonists

The diacidic amino acid L-glutamate is considered to be the endogenous agonist for cloned and *in situ* mGluRs. L-Glutamate is a flexible molecule that can adopt a number of low-energy conformations. This may account for its lack of selectivity for subpopulations of excitatory amino acid receptors. L-Glutamate is a highly potent agonist for all ionotropic glutamate receptors, including NMDA, AMPA, and kainate receptors. It also nonselectively activates all subtypes of mGluRs that have been cloned (*see* Nakanishi, 1992).

The endogenous amino acid L-aspartic acid is also an endogenous excitatory amino acid neurotransmitter candidate in certain synapses and has potent NMDA receptor agonist activity (Watkins et al, 1990). L-Aspartate can also weakly activate mGluRs. In rat hippocampal slices, it has been found to stimulate phosphoinositide hydrolysis (Schoepp and Johnson, 1989b) and inhibit forskolin-stimulated cAMP formation (Schoepp and Johnson, 1993b). The affinity of L-aspartic acid across the cloned mGluRs has not been determined, and its role as a potential endogenous mGluR agonist in mediating mGluR synaptic effects has not been investigated.

A number of other diacidic amino acid derivatives have been demonstrated to have mGluR agonist activities. In these molecules, an acidic bioisostere has replaced the carboxyl group of L-glutamate or L-aspartate. This has led to a number of compounds with significant mGluR activity, but generally has not provided mGluR receptor

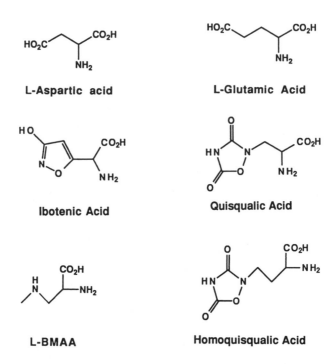

Fig. 1. Nonselective mGluR agonists. Compounds having both ionotropic and metabotropic glutamate receptor activities include: L-glutamic acid, L-aspartic acid, ibotenic acid, β-*N*-methylamino-L-alanine (BMAA), (S) quisqualate, and (S) homoquisqualate.

selectivity. For example, the compounds quisqualate and homo-quisqualate have a 1,2,4-oxadiazolidinedione substituted in place of the carboxy group of aspartic acid and glutamic acid, respectively (*see* Fig. 1). (S)Quisqualate (*see* Schoepp et al., 1992) and (S)homo-quisqualate (Porter et al., 1992b) are relatively potent agonists at phosphoinositide-coupled mGluRs. However, these compounds are also potent AMPA receptor agonists, and this greatly limits their use as pharmacological tools to study mGluR function. Nevertheless, quisqualate is one of the most potent mGluR agonists at certain mGluR subtypes, including phosphoinositide-coupled mGluR1α (Aramori and Nakanishi, 1992; Thomsen et al., 1993) and mGluR5 (Abe et al., 1992). In these clones, quisqualate acts as a full agonist with low-micromolar EC_{50} values. In contrast, the negatively coupled adenylyl cyclase-linked mGluRs, including rat mGluR2 (Tanabe et al., 1992), mGluR3 (Tanabe et al., 1993), mGluR4 (Tanabe et al., 1993), and mGluR6 (Nakajima et al., 1993), are relatively insensitive to quisqualate, although quisqualate does activate mGluR2 and mGluR3 with low potency. This is also true in rat brain slices where >100 μM quisqualate is needed to decrease forskolin-stimulated cAMP formation (Schoepp and Johnson, 1993b). Thus, it is useful to subclassify mGluRs into "quisqualate-sensitive" phosphoinositide-coupled mGluRs (mGluR1 and mGluR5) and "quisqualate-insensitive" negatively coupled adenylyl cyclase-linked mGluRs (mGluR2, mGluR3, mGluR4, and mGluR6). Metabotropic ^{3}H-glutamate binding sites that are quisqualate-sensitive and quisqualate-insensitive have been described, and these sites likely correspond to these subpopulations of mGluRs (Schoepp and True, 1992; Catania et al., 1993).

Ibotenic acid was one of the first recognized mGluR receptor agonists that stimulated phosphoinositide hydrolysis in brain slices (Nicoletti et al., 1986b) and cultured neurons (Nicoletti et al., 1986c; Manzoni et al., 1991a). This compound originated from a chemical series that contains the highly selective ionotropic agonist AMPA (Krogsgaard-Larsen et al., 1980) and has a 3-isoxazolone side chain at the distal acidic moiety (*see* Fig. 1). Ibotenate lacks AMPA receptor activity, but is a highly potent NMDA receptor agonist, in addition to having metabotropic activity. Whereas its NMDA receptor activities are potently antagonized by competitive and noncompetitive NMDA receptor antagonists, its metabotropic effects are not (*see* Schoepp et al., 1990a). When measuring phosphoinositide hydrolysis in rat hippocampal slices, ibotenate is a full agonist, but it has relatively low potency, requiring high-micromolar to millimolar concentrations to produce effects (Nicoletti et al., 1986b; Schoepp and Johnson, 1988). Ibotenate has no selectivity among subtypes of mGluRs, since it also has been found to decrease forskolin-stimulated cAMP formation in adult rat brain slices (Schoepp and Johnson, 1993a), stimulate cAMP formation in neonatal and adult rat brain slices (Casabono et al., 1992; Winder and Conn, 1992; Schoepp and Johnson, 1993b), and increase phospholipase D activity in rat hippocampal slices (Boss and Conn, 1992). These effects of ibotenate are likely mediated by metabotropic receptors, since they are not blocked by ionotropic glutamate receptor antagonists. Ibotenate also activates multiple cloned mGluRs with similar potency when compared to its *in situ* mGluR actions (high-micromolar to millimolar). This includes rat mGluR1α (Aramori and Nakanishi, 1992; Thomsen et al., 1993), mGluR2 (Tanabe et al., 1992), mGluR3 (Tanabe et al., 1993), mGluR4 (Tanabe et al., 1993), mGluR5 (Abe et al., 1992), and mGluR6 (Nakajima et al., 1993). Although ibotenate is one of the first recognized and most well-characterized mGluR agonists, its relatively low mGluR potency, its lack of selectivity for mGluR subtypes, and its potent NMDA agonist activity greatly limit its usefulness.

The compound β-*N*-methylamino-L-alanine (L-BMAA; *see* Fig. 1) is also an mGluR agonist, as illustrated by stimulation of phosphoinositide hydrolysis in rat brain slices and cultured cerebellar neurons (Copani et al., 1990, 1991; Manzoni et al., 1991b). However, this compound has mixed activity with effects on both NMDA and metabotropic receptors (Copani et al., 1991; Weiss et al., 1989). Interestingly, although L-BMAA possesses no distal acid group *per se* (*see* Fig. 1), it has been shown that its glutamate receptor agonist activities are dependent on bicarbonate ions, which may form an acidic carbonate adduct with L-BMAA (Weiss et al., 1989). It is interesting that L-BMAA is an exogenous excitotoxin, and its ingestion has been linked to the etiology of Guam amyotrophic lateral sclerosis-parkinsonism-dementia syndrome (*see* Meldrum and Garthwaite, 1990). The possible contribution of the mGluR activity in L-BMAA to this neurotoxic effect is not known.

Less-well-characterized endogenous mGluR agonists include the sulfur-containing amino acid derivatives L-cysteine sulfinic acid, L-cysteic acid, homocysteic acid, and homocysteine sulfinic acid. These compounds each stimulate phosphoinositide hydrolysis in neonatal rat brain slices with potencies comparable to L-glutamate and L-aspartate (Porter and Roberts, 1993), suggesting that they may act as endogenous mGluR (as well as iGluR) agonists in some synapses. L-Cysteine sulfinic acid has also been shown to activate

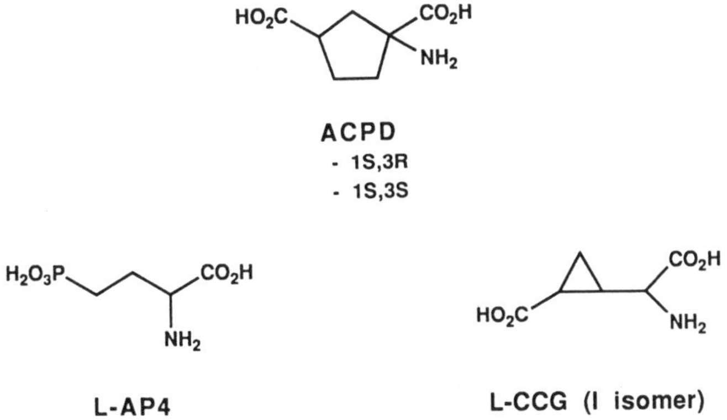

ACPD
- 1S,3R
- 1S,3S

L-AP4   L-CCG (I isomer)

Fig. 2. mGluR-selective agonists. Compounds that activate metabotropic glutamate receptors at concentrations having no appreciable effects on ionotropic glutamate receptors include: aminocyclopentane-1,3-dicarboxylic acid (ACPD, 1S,3R- and 1S,3S- isomers), L-2-amino-4-phosphonobutanoate (L-AP4), and α-carboxycyclopropylglycine (2S,3S,4S- isomer = L-CCG I).

mGluRs in the rat hippocampal slices that are coupled to enhanced phospholipase D activity and may be an endogenous agonist acting at the site in this tissue (Boss and Conn, 1993). The endogenous dipeptide *N*-acetylaspartylglutamate has been shown to decrease forskolin-stimulated cAMP formation selectively in cultured granule cells (Wroblewska et al., 1993), suggesting it may have a physiological role in activation of this class of mGluRs.

2.2. mGluR-Selective Agonists

Trans-ACPD was the first mGluR-selective agonist to be described, and the discovery of this compound greatly facilitated the investigation of the functional consequences of mGluR activation (*see* Section 1.3; Schoepp and Conn, 1993). ACPD is a conformationally constrained analog of L-glutamate (*see* Fig. 2) possessing four different stereoisomers (1S,3R, 1R,3S, 1S,3S, and 1R,3R). The 1S,3R and 1R,3S isomers of ACPD have also been termed "*trans*-ACPD," because the two carboxylic acid functionalities are disposed on opposite sides of the cyclopentane ring system. Likewise, the 1S,3S and 1R,3R isomers have been termed "*cis*-ACPD," since the carboxyl groups are oriented on the same face of the molecule. Racemic mixtures of these have thus been termed (±)-*trans*-ACPD (1SR,3RS-ACPD) and (±)-*cis*-ACPD (1SR,3SR-ACPD). This terminology is contrary to IUPAC rules and created some confusion (*see* Schoepp et al., 1990). Thus, it is suggested that any reference to ACPD isomers (pure or racemic) should adhere to the S and R absolute stereochemical designations, and the more ambiguous *"cis"* and *"trans"* terminology should be used with caution (*see* Schoepp et al., 1990a).

1S,3R-ACPD (or as *trans-ACPD*) produces a wide range of biochemical and cellular effects that are linked to the selective activation of mGluRs. 1S,3R-ACPD is a selective agonist for both phosphoinositide-coupled and cAMP-linked mGluRs, with micromolar potency (*see* Section 1.3., Schoepp and Conn, 1993). For example, in slices of rat hippocampus, the reported EC_{50} for 1S,3R-ACPD in increasing phosphoinositide hydrolysis is 19.4 μM. 1S,3R-ACPD also decreases forskolin-stimulated cAMP formation in this same preparation at similar concentrations (Schoepp et al., 1992). 1S,3R-ACPD appears to be a highly selective agonist for mGluRs. In rat brain membranes, 1S,3R-ACPD does not appreciably displace binding to NMDA (^{3}H-CGS 19755), AMPA (^{3}H-AMPA), or kainate (^{3}H-kainate) receptor sites at up to 1 mM (*see* Schoepp, 1993). In a functional assay of the rat hippocampal slice that measures ionotropic glutamate receptor activation by NMDA, AMPA, or kainate (enhanced ^{3}H-norepinephrine release), 1S,3R-ACPD produced no effect at up to 3 mM (Sacaan and Schoepp, 1992). Thus, based on the potency in activating mGluRs in the same preparation (rat hippocampus), it has >150-fold selectivity for activating mGluRs when compared to ionotropic glutamate receptors.

The potencies of the isomers of ACPD for cloned rat mGluRs have not been generally studied. However, racemic *trans*-ACPD activates both phosphoinositide coupled and cAMP-coupled mGluRs with potencies similar to reported *in situ* responses. In cells expressing rat mGluR1α and mGluR5 receptors, *trans*-ACPD increases phosphoinositide hydrolysis with EC_{50} values of 98 (Thomsen et al., 1993) and 10 μM (Abe et al., 1992), respectively. *Trans*-ACPD also activates rat mGluR2 and mGluR3 receptors as indexed by decreases in forskolin-stimulated cAMP formation with EC_{50} values of 5 μM (Tanabe et al., 1992) and 8 μM (Tanabe et al., 1993), respectively. Thus, it shows little selectivity among group one (mGluR1 and mGluR5) and group two (mGluR2 and mGluR3) mGluRs (*see* Nakanishi, 1992). However, the group three mGluRs, including mGluR4 (Tanabe et al., 1993) and mGluR6 (Nakajima et al., 1993), are relatively insensitive to *trans*-ACPD.

1S,3S-ACPD also has significant mGluR activity as demonstrated by its ability to activate phosphoinositide hydrolysis (Desai et al., 1992; Cartmell et al., 1993), inhibit forskolin-stimulated cAMP formation (Cartmell et al., 1993), potentiate-agonist-induced increases in cAMP formation (Cartmell et al., 1993; Winder et al., 1993), and activate phospholipase D activity in rat hippocampal slices (Boss and Conn, 1992). When measuring phosphoinositide hydrolysis and enhanced cAMP responses, 1S,3R-ACPD was more potent than 1S,3S-ACPD. However, 1S,3S-ACPD was more potent than 1S,3R-ACPD when measuring inhibition of forskolin-stimulated cAMP formation. This suggests that 1S,3S-ACPD may be a somewhat selective agonist for negatively coupled cAMP-linked mGluRs.

The compound L-AP4 is an analog of L-glutamate in which the distal acidic group has been replaced by a phosphonate moiety (*see* Fig. 2). Since the pK_a of phosphonic acid is <1, this molecule will be in a completely charged state at physiological pH. L-AP4 has been known for some time to act at presynaptic glutamate receptors in some synapses (i.e., perforant pathway) that function to decrease evoked release of glutamate (Koerner and Cotman, 1981) or in the retina, where it mimics the effects of L-glutamate by hyperpolarizing ON-bipolar cells (Miller and Slaughter, 1986) (for review, *see* Monaghan et al., 1989). The molecular basis for this effect of L-AP4 was unknown. It has now been recognized that L-AP4 has low-micromolar potency in activating certain mGluRs, including mGluR4 (Tanabe et al., 1993; Kristensen et al., 1993) and mGluR6 (Nakajima et al., 1993). Thus, L-AP4 may be considered a selective agonist for these mGluRs, and their *in situ* activation may be mediating L-AP4-sensitive receptor effects that have been defined electrophysiologically. Consistent with this, the aspartate analog L-serine-*O*-phosphate (L-SOP) also acts like L-AP4 to inhibit evoked release of glutamate from perforant path synapses (Ganong and Cotman, 1982). L-SOP is also a potent agonist in rat mGluR4- (Tanabe et al., 1993; Thomsen and Suzdak, 1993a) and mGluR6-expressing cells (Nakajima et al., 1993), but has no effect at mGluR1α (Thomsen and Suzdak, 1993a).

The 2-(carboxycyclopropyl)glycines (CCGs) are conformationally restrained analogs of glutamate that exist as eight isomers with various activities at both ionotropic and metabotropic glutamate receptors (Ohfune et al., 1993). In particular, (2S,3S,4S)α-(carboxy-cyclopropyl)glycine (L-CCG-I) is a relatively potent and selective agonist in activating mGluRs at concentrations having little effect on ionotropic glutamate receptors (*see* Shinozaki and Ishida, 1992). L-CCG I activates phosphoinositide hydrolysis in rat hippocampal synaptoneurosomes with an EC_{50} of 5 μM and was five times more potent than *trans*-ACPD in that tissue (Nakagawa et al., 1990). Similar concentrations of L-CCG-I elicit oscillatory currents in the *Xenopus* oocyte injected with rat brain mRNA and depolarize neonatal rat spinal motoneurons. These effects were not blocked by ionotropic glutamate receptor antagonists (Ishida et al., 1990). In cells expressing cloned rat mGluRs (Hayashi et al., 1992), L-CCG-I was the most potent CCG isomer in stimulating phosphoinositide hydrolysis at mGluR1a, with a EC_{50} of 50 μM. L-CCG-I was considerably more potent in mGluR2-expressing cells, and decreased forskolin-stimulated cAMP formation with an EC_{50} of 0.3 μM. This selectivity for negatively coupled cAMP-linked mGluRs has also been shown in rat striatal slices, where L-CCG-I decreases forskolin-stimulated cAMP formation with an EC_{50} of 0.9 μM, but its EC_{50} for stimulating phosphoinositide hydrolysis was 30 μM (Lombardi et al., 1993). Thus, L-CCG-I may a useful compound to probe further the cellular consequences of selectively activating negatively coupled cAMP-linked mGluRs *in situ*.

2.3. Miscellaneous mGluR Agonists

Compounds which act on mGluRs, but are less well characterized include 4-bromohomoibotenic acid (*see* Fig. 3), which has been reported to stimulate phosphoinositide hydrolysis in rat brain slices

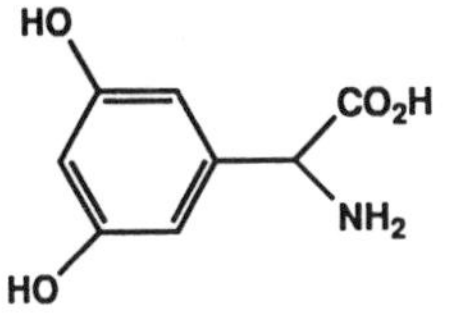

3,5-DHPG

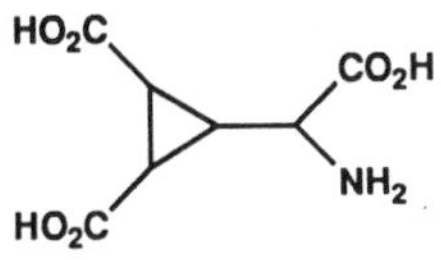

DCG-IV

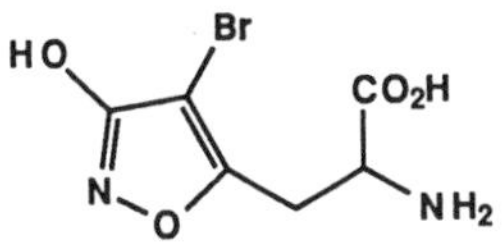

BrHomo Ibotenic Acid

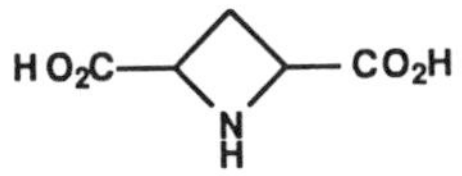

t-ADA

Fig. 3. Miscellaneous mGluR agonists. Less-well-characterized compounds with novel metabotropic glutamate receptor activities include: bromohomoibotenic acid (BrHomo Ibotenate), 3,5-dihydroxyphenylglycine (DHPG), (2,3-dicarboxycyclopropyl)glycine (2S,1'R,2'R,3'R-2-isomer = DCG-IV), and *trans*-azetidine-2,4-dicarboxylic acid (t-ADA).

(Chung et al., 1993). Interestingly, this compound produces additive effects when combined with the agonist 1S,3R-ACPD, suggesting that it is acting at a 1S,3R-ACPD-insensitive population of *in situ* phosphoinositide coupled mGluRs. Cloned rat mGluRs linked to phosphoinositide hydrolysis having such a pharmacology (1S,3R-ACPD-insensitive) have not yet been identified.

(D,L)3,5-Dihydroxyphenylglycine (*see* Fig. 3) was reported to be a more potent mGluR agonist than *trans*-ACPD in *Xenopus* oocytes expressing rat mGluR1α receptors (Ito et al., 1992). The selectivity of this compound for other mGluR subtypes and at ionotropic glutamate receptors was not reported.

(2S,1'R,2'R,3'R)-2-(2,3-Dicarboxycyclopropyl)glycine (DCG-IV) (*see* Fig. 3) is a triacidic amino acid derivative that has NMDA receptor agonist activity at micromolar concentrations (Ohfune et al., 1993). However, in the rat spinal motoneurons, preparation DCG-IV at 30 nM–1 μM concentration depressed monosynaptic excitation, suggesting a reduction in transmitter release from primary afferent nerve terminals (Ishida et al., 1993). Thus, it was more potent than L-CCG-I, 1S,3R-ACPD, and L-AP4 in that regard. DCG-IV has also been found to enhance quisqualate-stimulated phosphoinositide hydrolysis dramatically (EC$_{50}$ = 30 nM for this effect), while having no effect on basal hydrolysis (Nicoletti et al., 1993). Stimulation of phosphoinositide hydrolysis by 1S,3R-ACPD was not altered by DCG-IV. Thus, it was suggested that DCG-IV may act as a positive modulator of specific mGluRs in the rat hippocampus.

The simple amino acid derivative *trans*-azetidine-2,4-dicar-boxylic acid is an activator of mGluRs as demonstrated by stimulation of phosphoinositide hydrolysis in primary culture of rat cerebellar granule cells (Kozikowski et al., 1993). *Trans*-azetidine-2,4-dicarboxylic acid was less potent and less efficacious than *trans*-ACPD in this preparation, and was inactive when tested in rat mGluR1a expressing cells, suggesting that it may be an mGluR subtype-selective agonist.

3. mGluR Antagonist Pharmacology

3.1. Noncompetitive Inhibitors of mGluR Responses

L-2-Amino-3-phosphonopropanoate (L-AP3) and L-2-amino-4-phosphonobutanoate (L-AP4) are analogs of aspartate and glutamate, respectively, in which the distal carboxyl group is replaced by phosphonic acid (*see* Fig. 4). L-AP4, which is also a potent mGluR4 and mGluR6 agonist at low-micromolar concentration (*see* Section 2.2.) was the first reported inhibitor of mGluR-stimulated phosphoinositide hydrolysis. L-AP4 inhibited ibotenate activation of phosphoinositide hydrolysis in adult rat brain slices at 0.1–1 mm concentrations (Nicoletti et al., 1986b; Schoepp and Johnson, 1988). However, inhibition by L-AP4 was not found when using hippocampal slices from neonatal (6–7-old) rats (Nicoletti et al., 1986a; Schoepp and Johnson, 1989b). L-AP3 is more potent in this regard, and will inhibit ibotenate, quisqualate, L-glutamate, and *trans*-ACPD-induced increases in phosphoinositide hydrolysis in both adult and neonatal rat brain slices (IC$_{50}$ = 168 and 368 μM vs ibotenate in the

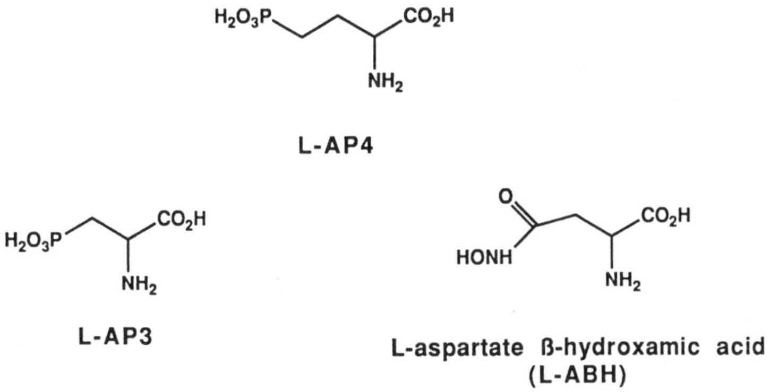

L-AP4

L-AP3

L-aspartate ß-hydroxamic acid
(L-ABH)

Fig. 4. Noncompetitive inhibitors of mGluRs. Compounds that have been found to inhibit mGluR responses in a noncompetitive fashion include L-2-amino-4-phosphonobutanoate (L-AP4), L-2-amino-3-phosphonopropanoate (L-AP3), and L-aspartate-β-hydroxamic acid (L-ABH).

adult and neonatal rat hippocampus, respectively) (Schoepp and Johnson, 1988, 1989a,b; Schoepp and Hillman, 1990; Schoepp et al., 1990b). L-AP3 has no appreciable ionotropic glutamate receptor affinity at up to 1 mM (Schoepp et al., 1990b). In rat brain slices, L-AP3 and L-AP4 are apparent noncompetitive inhibitors of mGluR-mediated phosphoinositide hydrolysis, producing long-lasting inhibition in vitro that is difficult to reverse (Schoepp et al., 1990b). Inhibition of mGluR-mediated phosphoinositide hydrolysis by L-AP3 was shown to be dependent on the presence of extracellular calcium (Lonart et al., 1992). L-AP3 and L-AP4 also exhibit partial agonist characteristics by modestly stimulating phosphoinositide hydrolysis *per se*, and this limits the degree of inhibition attainable against a more efficacious agonist (*see* Schoepp et al., 1990a). Selective inhibition of ibotenate-, quisqualate-, and *trans*-ACPD-induced phosphoinositide hydrolysis can also be produced *ex vivo* by the parenteral administration of D,L-AP3 (250 or 500 mg/kg ip) to neonatal rats (Schoepp and Johnson, 1991). These data suggest that in some tissues, L-AP3 may be acting as a desensitizing agonist at phosphoinositide-coupled mGluRs, and in this manner, subsequent responses to other agonists, such as *trans*-ACPD, quisqualate, and ibotenate, are reduced.

In cultured murine striatal neurons, L-AP3 also acts as a weak agonist in stimulating phosphoinositide hydrolysis, but it inhibits quisqualate-induced phosphoinositide hydrolysis in a competitive manner (Manzoni et al., 1991a). Also, when studying mGluR-mediated currents in the *Xenopus* oocyte, L-AP3 acts as a readily reversible antagonist with no intrinsic activity *per se* (Tanabe et al., 1991). Work in the *Xenopus* oocyte suggests that L-AP3 is an mGluR subtype-selective antagonist, since it is a more effective inhibitor of *trans*-ACPD-induced currents when compared to quisqualate (Tanabe et al., 1991) or 3,5-dihydroxyphenylglycine (Ito et al., 1992).

In rat mGluR1α-expressing cells, L-AP3 is a relatively weak antagonist and only partially inhibits (about 44%) agonist-induced phophoinositide hydrolysis at 1 mM (Thomsen et al., 1993). Other cloned mGluRs, such as rat mGluR5, are insensitive to L-AP3 or L-AP4, even at 1 mM (Abe et al., 1992).

The usefulness of L-AP3 as a selective antagonist for studies of mGluR function is limited by its low potency and not well-understood mechanism of inhibition in tissues where it acts noncompetitively. More recent studies have shown that L-AP3 can also alter other mGluR-mediated receptor responses in rat brain slices. It blocks *trans*-ACPD-induced increases in cAMP formation in the neonatal rat hippocampus (Casabona et al., 1992; Winder and Conn, 1992; Schoepp and Johnson, 1993b), and mimics *trans*-ACPD in the adult rat hippocampus by decreasing forskolin-stimulated cAMP formation (Schoepp and Johnson, 1993a) and activating phospholipase D (Boss and Conn, 1992). Thus, the inhibitory effects of L-AP3 on mGluR functional responses, such as induction of long-term potentiation (*see* Anwyl, 1991), could be the result of actions on any or all of these second-messenger systems.

The aspartate analog L-aspartate-β-hydroxamate is also a noncompetitive and long-lasting inhibitor of mGluR agonist-mediated phosphoinositide hydrolysis in rat brain slices, with a potency about ten times greater than L-AP3 (Ormandy, 1992; Porter et al., 1992a; Littman et al., 1993). It is also a highly potent glutamate uptake inhibitor at similar concentrations (Littman et al., 1993). Effects on cloned mGluRs and other second-messenger responses are not known.

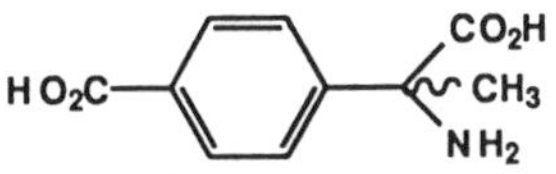

(+) α-Methyl-4-Carboxy PG

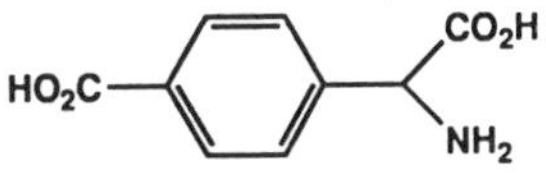

(S) 4-Carboxy PG

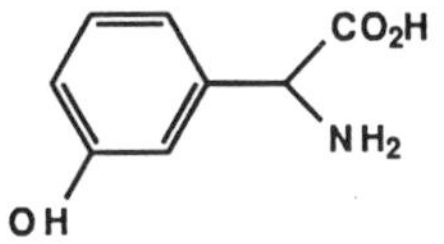

(S) 3-Hydroxy PG

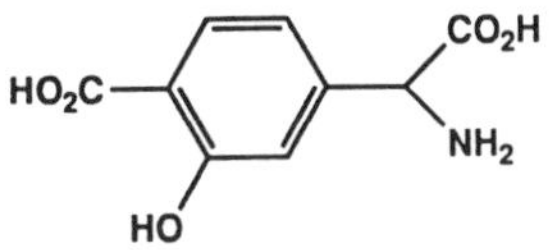

(S) 4-Carboxy-3-hydroxy PG

Fig. 5. Competitive mGluR antagonists include: the phenylglycine derivatives (S)-4-carboxyphenylglycine (4-carboxy PG), (S)-3-hydroxyphenyglycine (3-hydroxy PG), (S)-4-carboxy-3-hydroxyphenylglycine (4-carboxy-3-hydroxy PG), and (+)-α-methyl-4-carboxyphenylglycine (α-methyl-4-carboxy PG).

3.2. Competitive mGluR Antagonists

A series of phenylglycine derivatives having partial agonist and/or competitive antagonist properties against mGluR-mediated second-messenger and electrophysiological responses have been described. These include (S)-4-carboxyphenylglycine, (S)-3-hydroxyphenyglycine, (S)-4-carboxy-3-hydroxyphenylglycine, and (+)-α-methyl-4-carboxyphenylglycine (*see* Fig. 5). In rat cerebral cortex slices, rightward shifts in the *trans*-ACPD concentration–effect curve for increased phosphoinositide hydrolysis were produced by (RS)-4-carboxyphenylglycine and (RS)-α-methyl-4-carboxyphenylglycine with K_B values of 1.7 and 1.2 mM, respectively (Eaton et al., 1993). This activity was reported to reside in the S-isomers of these compounds (Eaton et al., 1993). Using cells expressing rat mGluR1α, (RS)-4-carboxy-3-hydroxyphenylglycine competitively inhibited glutamate-induced phosphoinositide hydrolysis with a K_B of 29 μM (Thomsen and Suzdak, 1993b). In neonatal rat cerebral cortical slices, (S)-3-hydroxyphenyglycine and (S)-4-carboxy-3-hydroxyphenylglycine were shown to be weak agonists that stimulate phosphoinositide hydrolysis with lower potency and efficacy than *trans*-ACPD (Birse et al., 1993). Nevertheless, depolarization of neonatal rat motoneurons evoked by *trans*-ACPD is antagonized by (S)-4-carboxyphenyl-glycine, (S)-3-hydroxyphenyglycine, and (S)-4-carboxy-3-hydroxy-phenylglycine at 500–1000 μ*M* concentrations (Birse et al., 1993). *Trans*-ACPD-induced depolarization of the neonatal rat motoneurons and ionotophoretic applied *trans*-ACPD excitation of thalamic neurons were shown to be competitively antagonized by (+)-α-methyl-4-carboxyphenylglycine with an apparent K_B of 144 μ*M* (Jane et al., 1993). At 250 μ*M*, (+)-α-methyl-4-carboxyphenylglycine did not antagonize the depolarizing effects of NMDA and AMPA. Thus, (+)-α-methyl-4-carboxyphenylglycine appears to be the most potent and selective mGluR antagonist from this series, and has been useful as a functional antagonist to define the role of mGluRs in physiological processes, such as long-term potentiation (Bashir et al., 1993).

4. Future mGluR Pharmacology

The recognition of a new class of glutamate receptors that are "metabotropic" in character has now been clearly established by both molecular and pharmacological means. These receptors appear to be highly hetergeneous at the molecular level, and current understanding of their functions is limited. The pharmacology of this area is in its early stages. However, the availability of cloned mGluRs should facilitate the discovery of more selective pharmacological agents. The discovery of selective agonists and antagonists for the subtypes of these receptors will be the likely focus in the future. Subtype-selective mGluR agonists and antagonists will serve as highly useful research tools to study the functions of specific mGluRs in normal and pathological states. This may also lead to highly novel approaches to treat neurological and psychiatric diseases involving altered glutamatergic neuronal transmission in select areas of the CNS.

Acknowledgments

Thanks are given to James A. Monn and Paul L. Ornstein for their helpful suggestions during the preparation of this chapter.

References

Abe, T., Sugihara, H., Nawa, H., Shigemoto, R., Mizuno, N., and Nakanishi, S. (1992) Molecular characterization of a novel metabotropic glutamate receptor mGluR5 coupled to inositol phosphate/Ca^{2+} signal transduction. *J. Biol. Chem.* **267,** 13,361–13,368.

Ahmed, Z., Lewis, C. A., and Faber, D. S. (1990) Glutamate stimulates release of Ca^{2+} from internal stores in astroglia. *Brain Res.* **516,** 165–169.

Anwyl, R. (1991) The role of the metabotropic receptor in synaptic plasticity. *Trends Pharmacol. Sci.* **12,** 324–326.

Aramori, I. and Nakanishi, S. (1992) Signal transduction and pharmacological characteristics of a metabotropic glutamate receptor, mGluR1, in transfected CHO cell. *Neuron* **8,** 757–765.

Aronica, E., Nicoletti, F., Condorelli, D. F., and Balazs, R. (1993) Pharmacological characteristics of metabotropic glutamate receptors in cultured cerebellar granule cells. *Neurochem. Res.* **18,** 605–612.

Bardsley, M. E. and Roberts, P. J. (1983) Stimulation of phosphatidylinositol turnover in rat brain by glutamate and aspartate. *Br. J. Pharmacol.* **79,** 401P.

Bashir, Z. I., Bortolotto, Z. A., Davies, C. H., Berretta, N., Irving, A. J., Seal, A. J., Henley, J. M., Jane, D. E., Watkins, J. C., and Collingridge, G. L. (1993) Induction of LTP in the hippocampus needs synaptic activation of glutamate metabotropic receptors. *Nature* **363,** 347–350.

Berridge, M. J. and Irvine, R. F. (1984) Inositol trisphosphate, a novel second messenger in cellular signal transduction. *Nature* **312,** 315–321.

Berridge, M. J., Downes, C. P., and Hanley, M. R. (1982) Lithium amplifies agonist-dependent phosphatidylinositol responses in brain and salivary glands. *Biochem. J.* **206,** 587–595.

Birse, E. F., Eaton, S. A., Jane, D. E., Jones P. L. St. J., Porter, R. H. P., Pook, P. C.-K., Sunter, D. C., Udvarhelyi, P. M., Wharton, B., Roberts, P. J., Salt, T. E., and Watkins, J. C. (1993) Phenylglycine derivatives as new pharmacological tools for investigating the role of metabotropic glutamate receptors in the central nervous system. *Neuroscience* **52,** 481–488.

Boss, V. and Conn, P. J. (1992) Metabotropic excitatory amino acid receptor activation stimulates phospholipase D in hippocampal slices. *J. Neurochem.* **59,** 2340–2343.

Boss, V. and Conn, P. J. (1993) L-Cysteine sulfinic acid (L-CSA): an endogenous agonist of a metabotropic excitatory amino acid receptor subtype. *Abstr. Soc. Neurosci.* **19,** 626.

Boss, V., Desai, M. A., Smith, T. S., and Conn, P. J. (1992) *Trans*-ACPD-induced phosphoinositide hydrolysis and modulation of hippocampal pyramidal cell excitability do not undergo parallel developmental regulation. *Brain Res.* **594,** 181–188.

Cartmell, J., Kemp, J. A., Alexander, S. P. H., Hill, S. J., and Kendall, D. A. (1992) Inhibition of forskolin-stimulated cyclic AMP formation by 1-aminocyclopentane-*trans*-1,3-dicarboxylate in guinea-pig cerebral cortical slices. *J. Neurochem.* **58,** 1964–1966.

Cartmell, J., Curtis, A. R., Kemp, J. A., Kendall, D. A., and Alexander, S. P. H. (1993) Subtypes of metabotropic excitatory amino acid receptor distinguished by stereoisomers of the rigid glutamate analogue, 1-aminocyclopentane-1,3-dicarboxylic acid. *Neurosci. Lett.* **153,** 107–110.

Casabona, G., Genazzani, A. A., Di Stefano, M., Sortino, M. A., and Nicoletti, F. (1992) Developmental changes in the modulation of cyclic AMP formation by the metabotropic glutamate receptor agonist 1S,3R-aminocyclopentane-1,3-dicarboxylic acid in brain slices. *J. Neurochem.* **59,** 1161–1163.

Catania, M. V., Hollingsworth, Z., Penney, J. B., and Young, A. B. (1993) Quisqualate resolves two distinct metabotropic [^{3}H]glutamate binding sites. *NeuroReport* **4,** 311–313.

Chung, D. S., Winder, D. G., and Conn, P. J. (1993) 4-Bromohomoibotenic acid selectively activates an ACPD-insensitive metabotropic glutamate receptor coupled to phosphoinositide hydrolysis in rat cortical slices. *J. Neurochem.* in press.

Copani, A., Canonico, P. L., and Nicoletti, F. (1990) β-N-Methylamino-L-alanine (L-BMAA) is a potent agonist of "metabolotropic" glutamate receptors. *Eur. J. Pharmacol.* **181,** 327–328.

Copani, A., Canonico, P. L., Catania, M. V., Aronica, E., Bruno, V., Ratti, E., van Amsterdam, F. T. M., Gaviraghi, G., and Nicoletti, F. (1991) Interaction between β-N-methylamino-L-alanine and excitatory amino acid receptors in brain slices and neuronal cultures. *Brain Res.* **558,** 79–86.

Desai, M. A. and Conn, P. J. (1990) Selective activation of phosphoinositide hydrolysis by a rigid analogue of glutamate. *Neurosci. Lett.* **109,** 157–162.

Desai, M. A., Smith, T. S., and Conn, P. J. (1992) Multiple metabotropic glutamate receptors regulate hippocampal function. *Synapse* **12,** 206–213.

Eaton, S. A., Jane, D. E., Jones, P. L. St. J., Porter, R. H. P., Pook, P. C.-K., Sunter, D. C., Udvarhelyi, P. M., Roberts, P. J., Salt, T. E., and Watkins, J. C. (1993) Competitive antagonism at metabotropic glutamate receptors by (S)-4-carboxyphenylglycine and (RS)-α-methyl-4-carboxyphenylglycine. *Eur. J. Pharmacol.* **244,** 195–197.

Fisher, S. K. and Agranoff, B. W. (1987) Receptor activation and inositol lipid hydrolysis in neural tissues. *J. Neurochem.* **48,** 999–1017.

Ganong, A. H. and Cotman, C. W. (1982) Acidic amino acid antagonists of lateral perforant path synaptic transmission: agonist-antagonist interactions in the dentate gyrus. *Neurosci. Lett.* **34,** 195–200.

Gerber, U., Sim J. A., and Gahwiler, B. H. (1992) Reduction of potassium conductances mediated by metabotropic glutamate receptors in rat CA3 pyramidal cells does not require protein kinase C or protein kinase A. *Eur. J. Neurosci.* **4,** 792–797.

Hallcher, L. M. and Sherman, W. R. (1980) The effects of lithium ions and other agents on the activity of *myo*-inositol-1-phosphatase from bovine brain. *J. Biol. Chem.* **255,** 10,896–10,901.

Hayashi, Y., Tanabe, Y., Aramori, I., Masu, M., Shimamoto, K., Ohfune, Y., and Nakanishi, S. (1992) Agonist analysis of 2-(carboxycyclopropyl)glycine isomers for cloned metabotropic glutamate receptor subtypes expressed in Chinese hamster ovary cells. *Br. J. Pharmacol.* **107,** 539–543.

Holler, T., Cappel, E., Klein, J., and Löffelholz, K. (1993) Glutamate activates phospholipase D in hippocampal slices of newborn and adult rats. *J. Neurochem.* **61,** 1569–1572.

Holzwarth, J. A., Gibbons, S. J., Brorson, J. R., Philipson, L. H., and Miller, R. J. (1993) Glutamate receptor agonists stimulate diverse calcium responses in different types of cultured rat cortical glial cells. *J. Neurosci.* in press.

Honore, T., Davies, S. N., Drejer, J., Fletcher, E. J., Jacobsen, P., Lodge, D., and Nielsen, F. E. (1988) Quinoxalinediones: potent competitive non-NMDA glutamate receptor antagonists. *Science* **241,** 701–703.

Houamed, K. M., Kuijper, J. L., Gilbert, T. L., Haldeman, B. A., O'Hara, P. J., Mulvihill, E. R., Almers, W., and Hagen, F. S. (1991) Cloning, expression, and gene structure of a G protein-coupled glutamate receptor from the rat brain. *Science* **252,** 1318–1321.

Irving, A. J., Collingridge, G. L., and Schofield, J. G. (1992) Interactions between Ca^{2+} mobilizing mechanisms in cultured rat cerebellar granule cells. *J. Physiol.* **456,** 667–680.

Irving, A. J., Schofield, J. G., Watkins, J. C., Sunter, D. C., and Collingridge, G. L. (1990) 1S,3R-ACPD stimulates and L-AP3 blocks Ca^{2+} mobilization in rat cerebellar neurons. *Eur. J. Pharmacol.* **186,** 363–365.

Ishida, M., Akagi, H., Shimamoto, K., Ohfune, Y., and Shinozaki, H. (1990) A potent metabotropic glutamate receptor agonist: electrophysiological actions of a conformationally restricted glutamate analogue in the rat spinal cord and *Xenopus* oocytes. *Brain Res.* **537,** 311–314.

Ishida, M., Saitoh, T., Shimamoto, K., Ohfune, Y., and Shinozaki, H. (1993) A novel metabotropic glutamate receptor agonist: marked depression of monosynaptic excitation in the newborn rat isolated spinal cord. *Br. J. Pharmacol.* **109,** 1169–1177.

Ito, I., Kohda, A., Tanabe, S., Hirose, E., Hayashi, M., Mitsunaga, S., and Sugiyama, H. (1992) 3,5-Dihydroxyphenyl-glycine: A potent agonist of metabotropic glutamate receptors. *NeuroReport* **3,** 1013–1016.

Jane, D. E., Jones, P. L. St. J., Pook, P. C.-K., Salt, T. E., Sunter, D. C., and Watkins, J. C. (1993) Stereospecific antagonism by (+)-α-methyl-4-carboxcyphenylglycine (MCPG) of (1S,3R)-ACPD-induced effects in neonatal rat motoneurones and rat thalamic neurones. *Neuropharmacology* **32,** 725–727.

Koerner, J. F. and Cotman, C. W. (1981) Micromolar L-2-amino-4-phosphonobutyric acid selectively inhibits perforant path synapses from lateral entorhinal cortex. *Brain Res.* **216,** 192–198.

Kozikowski, A. P., Tuckmantel, W., Liao, Y., Manev, H., Ikonomovic, S., and Wroblewski, J. T. (1993) Synthesis and metabotropic receptor activity of the novel rigidified glutamate analogues (+)- and (–)-*trans*-azetidine-2,4-dicarboxylic acid and their *N*-methyl derivatives. *J. Med. Chem.* **36,** 2706–2708.

Kristensen, P., Suzdak, P. D., and Thomsen, C. (1993) Expression pattern and pharmacology of the rat type IV metabotropic glutamate receptor. *Neurosci. Lett.* **155,** 159–162.

Krogsgaard-Larsen, P., Honore, T., Hansen, J. J., Curtis, D. R., and Lodge, D. (1980) New class of glutamate agonist structurally related to ibotenic acid. *Nature* **284,** 64–66.

Littman, L., Glatt, B. S., and Robinson, M. B. (1993) Multiple subtypes of excitatory amino acid receptors coupled to the hydrolysis of phosphoinositides in rat brain. *J. Neurochem.* **61,** 586–593.

Lombardi, G., Alesiani, M., Leonardi, P., Cherici, G., Pellicciari, R., and Moroni, F. (1993) Pharmacological characterization of the metabotropic glutamate receptor inhibiting D-[^{3}H]aspartate output in rat striatum. *Brit. J. Pharmacol.* **110,** 1407–1412

Lonart, G., Alagarsamy, S., Ravula, R., Wang, J., and Johnson, K. M. (1992) Inhibition of the phospholipase C-linked metabotropic glutamate receptor by 2-amino-3-phosphonopropionate is dependent on extracellular calcium. *J. Neurochem.* **59,** 772–775.

Manzoni, O., Fagni, L., Pin, J.-P., Rassendren, F., Poulat, F., Sladeczek, F., and Bockaert, J. (1990) *(trans)*-1-Amino-cyclopentyl-1,3-dicarboxylate stimulates quisqualate phosphoinositide-coupled receptors but not ionotropic glutamate receptors in striatal neurons and *xenopus* oocytes. *Mol. Pharmacol.* **38,** 1–6.

Manzoni, O. J. J., Poulat, F., Do, E., Sahuquet, A., Sassetti, I., Bockaert, J., and Sladeczek, F. A. J. (1991a) Pharmacological characterization of the quisqualate receptor coupled to phospholipase C (Qp) in striatal neurons. *Eur. J. Pharmacol.* **207,** 231–241.

Manzoni, O. J. J., Prezeau, L., and Bockaert, J. (1991b) β-*N*-methylamino-L-alanine is a low-affinity agonist of metabotropic glutamate receptors. *NeuroReport* **2,** 609–611.

Masu, M., Tanabe, Y., Tsuchida, K., Shigemoto, R., and Nakanishi, S. (1991) Sequence and expression of a metabotropic glutamate receptor. *Nature* **349,** 760–765.

Mayer, M. L. and Miller, R. J. (1990) Excitatory amino acid receptors, second messengers and regulation of intracellular Ca^{2+} in mammalian neurons. *Trends Pharmacol. Sci.* **11,** 254–260.

Meldrum, B. and Garthwaite, J. (1990) Excitatory amino acid neurotoxicity and neurodegenerative disease. *Trends Pharmacol. Sci.* **11,** 379–387.

Miller, R. F. and Slaughter, M. M. (1986) Excitatory amino acid receptors of the retina: diversity of subtypes and conductance mechanisms. *Trends Neurosci. (May),* 211–218.

Minakami, R., Katsuki, F., and Sugiyama, H. (1993) A variant of metabotropic glutamate receptor subtype 5: an evolutionarily conserved insertion with no termination codon. *Biochem. Biophys. Res. Comm.* **194,** 622–627.

Monaghan, D. T., Bridges, R. J., and Cotman, C. W. (1989) The excitatory amino acid receptors: their classes, pharmacology, and distinct properties in the function of the central nervous system. *Annu. Rev. Pharmacol. Toxicol.* **29,** 365–402.

Murphy, S. N. and Miller, R. J. (1988) A glutamate receptor regulates Ca^{2+} mobilization in hippocampal neurons. *Proc. Natl. Acad. Sci. USA* **85,** 8737–8741.

Nakagawa, Y., Saitoh, K., Ishihara, T., Ishida, M., and Shinozaki, H. (1990) (2S,3S,4S)α-(Carboxycyclopropyl)glycine is a novel agonist of metabotropic glutamate receptors. *Eur. J. Pharmacol.* **184,** 205–206.

Nakajima, Y., Iwakabe, H., Akazawa, C., Nawa, H., Shigemoto, R., Mizuno, N., and Nakanishi, S. (1993) Molecular characterization of a novel retinal metabotropic glutamate receptor mGluR6 with a high agonist selectivity for L-2-amino-4-phosphonobutyrate. *J. Biol. Chem.* **266**, 11,868–11,873.

Nakanishi, S. (1992) Molecular diversity of glutamate receptors and implications for brain function. *Science* **258**, 597–603.

Nicoletti, F., Casabona, G., Genazzani, A. A., L'Episcopo, M. R., and Shinozaki, H. (1993) (2*S*,1'*R*,2'*R*,3'*R*)-2-(2,3-Dicarboxycyclopropyl)glycine enhances quisqualate-stimulated inositol phospholipid hydrolysis in hippocampal slices. *Eur. J. Pharmacol.* **245**, 297–298.

Nicoletti, F., Iadarola, M. J., Wroblewski, J. T., and Costa, E. (1986a) Excitatory amino acid recognition sites coupled with inositol phospholipid metabolism: Developmental changes and interaction with α_1-adrenoceptors. *Proc. Natl. Acad. Sci. USA* **83**, 1931–1935.

Nicoletti, F., Meek, J. L., Iadarola, M. J., Chuang, D. M., Roth, B. L., and Costa, E. (1986b) Coupling of inositol phospholipid metabolism with excitatory amino acid recognition sites in rat hippocampus. *J. Neurochem.* **46**, 40–46.

Nicoletti, F., Wroblewski, J. T., Novelli, A., Alho, H., Guidotti, A., and Costa, E. (1986c) The activation of inositol phospholipid metabolism as a signal-transducing system for excitatory amino acids in primary cultures of cerebellar granule cells. *J. Neurosci.* **6**, 1905–1911.

Nishizuka, Y. (1984) Turnover of inositol phospholipids and signal transduction. *Science* **225**, 1365–1370.

Ohfune, Y., Shimamoto, K., Ishida, M., and Shinozaki, H. (1993) Synthesis of L-2-(2,3-dicarboxycyclopropyl)glycines, novel conformationally restricted glutamate analogues. *Bioorganic & Med. Chem. Lett.* **3**, 15–18.

Okada, D. (1992) Two pathways of cyclic GMP production through glutamate receptor-mediated nitric oxide synthesis. *J. Neurochem.* **59**, 1203-1210.

Ormandy G. C. (1992) Inhibition of excitatory amino acid-stimulated phosphoinositide hydrolysis in rat hippocampus by L-aspartate-β-hydroxamate. *Brain Res.* **572**, 103–107.

Osborne N. N. (1990) Stimulatory and inhibitory actions of excitatory amino acids on inositol phospholipid metabolism in rabbit retina. Evidence for a specific quisqualate receptor subtype associated with neurones. *Exp. Eye Res.* **50**, 397–405.

Palmer, E., Monaghan, D. T., and Cotman, C. W. (1988) Glutamate receptors and phosphoinositide metabolism: stimulation via quisqualate receptors is inhibited by *N*-methyl-D-aspartate receptor activation. *Mol. Brain Res.* **4**, 161–165.

Palmer, E., Monaghan, D. T., and Cotman, C. W. (1989) Trans-ACPD, a selective agonist of the phosphoinositide-coupled excitatory amino acid receptor. *Eur. J. Pharmacol.* **166**, 585–587.

Pearce, B., Albrecht, J., Morrow, C., and Murphy, S. (1986) Astrocyte glutamate receptor activation promotes inositol phospholipid turnover and calcium flux. *Neurosci. Lett.* **72**, 335–340.

Porter, R. H. P. and Roberts, P. J. (1993) Glutamate metabotropic receptor activation in neonatal rat cerebral cortex by sulphur-containing excitatory amino acids. *Neurosci. Lett.* **154**, 78–80.

Porter, R. H. P., Briggs, R. S. J., and Roberts, P. J. (1992a) L-Aspartate-β-hydroxy-amate exhibits mixed agonist/antagonist activity at the glutamate metabotropic receptor in rat neonatal cerebrocortical slices. *Neurosci. Lett.* **144,** 87–89.

Porter, R. H. P., Roberts, P. J., Jane, D. E., and Watkins, J. C. (1992b) (S)homoquisqualate: a potent agonist at the glutamate metabotropic receptor. *Br. J. Pharmacol.* **106,** 509–510.

Recasens, M., Guiramand, J., Nourigat, A., Sassetti, I., and Devilliers, G. (1988) A new quisqualate receptor subtype (sAA$_2$) responsible for the glutamate-induced inositol phosphate formation in rat brain synaptoneurosomes. *Neurochem. Int.* **13,** 463–467.

Recasens, M., Sassetti, I., Nourigat, A., Sladeczek, F., and Bockaert, J. (1987) Characterization of subtypes of excitatory amino acid receptors involved in the stimulation of inositol phosphate synthesis in rat brain synaptoneurosomes. *Eur. J. Pharmacol.* **141,** 87–93.

Sacaan, A. I. and Schoepp, D. D. (1992) Activation of hippocampal metabotropic excitatory amino acid receptors leads to seizures and neuronal damage. *Neurosci. Lett.* **139,** 77–82.

Schoepp, D. D. (1993) The biochemical pharmacology of metabotropic glutamate receptors. *Biochem. Soc. Trans.* **21,** 97–102.

Schoepp, D. D. and Conn, P. J. (1993) Metabotropic glutamate receptors in brain function and pathology. *Trends in Pharmacol. Sci.* **14,** 13–20.

Schoepp, D. D. and Hillman, C. C. (1990) Developmental and pharmacological characterization of quisqualate, ibotenate, and *trans*-1-amino-1,3-cyclopentanedicarboxylic acid stimulations of phosphoinositide hydrolysis in rat cortical brain slices. *Biogenic Amines* **7,** 331–340.

Schoepp, D. D. and Johnson, B. G. (1988) Excitatory amino acid agonist-antagonist interactions at 2-amino-4-phosphonobutyric acid-sensitive quisqualate receptors coupled to phosphoinositide hydrolysis in slices of rat hippocampus. *J. Neurochem.* **50,** 1605–1613.

Schoepp, D. D. and Johnson B. G. (1989a) Comparison of excitatory amino acid-stimulated phosphoinositide hydrolysis and *N*-[³H]acetylaspartylglutamate binding in rat brain: Selective inhibition of phosphoinositide hydrolysis by 2-amino-3-phosphonopropionate. *J. Neurochem.* **53,** 273–278.

Schoepp, D. D. and Johnson, B. G. (1989b) Inhibition of excitatory amino acid-stimulated phosphoinositide hydrolysis in the neonatal rat hippocampus by 2-amino-3-phosphonopropionate. *J. Neurochem.* **53,** 1865–1870.

Schoepp, D. D. and Johnson, B. G. (1991) *In vivo* 2-amino-3-phosphonopropionic acid administration to neonatal rats selectively inhibits metabotropic excitatory amino acid receptors *ex vivo* in brain slices. *Neurochem. Int.* **18,** 411–417.

Schoepp, D. D. and Johnson, B. G. (1993a) Pharmacology of metabotropic glutamate receptor inhibition of cyclic AMP formation in the adult rat hippocampus. *Neurochem. Int.* **22,** 277–283.

Schoepp, D. D. and Johnson, B. G. (1993b) Metabotropic glutamate receptor modulation of cAMP accumulation in the neonatal rat hippocampus. *Neuropharmacology* **32,** 1359–1365.

Schoepp, D. D. and True, R. A. (1992) 1*S*,3*R*-ACPD-sensitive (metabotropic) [³H]glutamate receptor binding in membranes. *Neurosci. Lett.* **145,** 100–104.

Schoepp, D., Bockaert, J., and Sladeczek, F. (1990a) Pharmacological and functional characteristics of metabotropic excitatory amino acid receptors. *Trends in Pharmacol. Sci.* **11,** 508–515.

Schoepp, D. D., Johnson, B. G., Smith, E. C. R., and McQuaid, L. A. (1990b) Stereoselectivity and mode of inhibition of phosphoinositide-coupled excitatory amino acid receptors by 2-amino-3-phosphonopropionic acid. *Molecular Pharmacol.* **38,** 222–228.

Schoepp, D. D., Johnson, B. G., Salhoff, C. R., McDonald, J. W., and Johnston, M. V. (1991a) In vitro and in vivo pharmacology of *trans*- and *cis*-(±)-1-amino-1,3-cyclopentanedicarboxylic acid: dissociation of metabotropic and ionotropic excitatory amino acid receptor effects. *J. Neurochem.* **56,** 1789–1796.

Schoepp, D. D., Johnson, B. G., True, R. A., and Monn, J. A. (1991b) Comparison of (1*S*,3*R*)-1-aminocyclopentane-1,3-dicarboxylic acid (1*S*,3*R*-ACPD)- and 1*R*,3*S*-ACPD-stimulated brain phosphoinositide hydrolysis. *Eur. J. Pharmacol.—Mol. Pharmacol. Section* **207,** 351–353.

Schoepp, D. D., Johnson, B. G., and Monn, J. A. (1992) Inhibition of cyclic AMP formation by a selective metabotropic glutamate receptor agonist. *J. Neurochem.* **58,** 1184–1186.

Shinozaki, H., and Ishida, M. (1992) A metabotropic L-glutamate receptor agonist: pharmacological difference between rat central neurones and crayfish neuromuscular junctions. *Comp. Biochem. Physiol.* **103C,** 13–17.

Sladeczek, F., Pin, J.-P., Recasens, M., Bockaert, J., and Weiss, S. (1985) Glutamate stimulates inositol phosphate formation in striatal neurones. *Nature* **317,** 717–719.

Sugiyama, H., Ito, I., and Hirono, C. (1987) A new type of glutamate receptor linked to inositol phospholipid metabolism. *Nature* **325,** 531–533.

Sugiyama, H., Ito, I., and Watanabe, M. (1989) Glutamate receptor subtypes may be classified into two major categories: a study on xenopus oocytes injected with rat brain mRNA. *Neuron* **3,** 129–132.

Tanabe, S., Ito, I., and Sugiyama, H. (1991) Possible heterogeneity of metabotropic glutamate receptors induced in *xenopus* oocytes by rat brain mRNA. *Neurosci. Res.* **10,** 71–77.

Tanabe, Y., Masu, M., Ishii, T., Shigemoto, R., and Nakanishi, S. (1992) A family of metabotropic glutamate receptors. *Neuron* **8,** 169–179.

Tanabe, Y., Nomura, A., Masu, M., Shigemoto, R., Mizuno, N., and Nakanishi, S. (1993) Signal transduction, pharmacological properties, and expression patterns of two rat metabotropic glutamate receptors, mGluR3 and mGluR4. *J. Neurosci.* **13,** 1372–1378.

Thomsen, C. and Suzdak, P. D. (1993a) 4-Carboxy-3-hydroxyphenylglycine, an antagonist at type I metabotropic glutamate receptors. *Eur. J. Pharmacol.* **245,** 299–301.

Thomsen, C. and Suzdak, P. D. (1993b) Serine-*O*-phosphate has affinity for type-IV, but not type-I, metabotropic glutamate receptor. *NeuroReport* **4,** 1099–1101.

Thomsen, C., Mulvihill, E. R., Haldeman, B., Pickering, D. S., Hampson, D. R., and Suzdak, P. D. (1993) A pharmacological characterization of the mGluR1α subtype of the metabotropic glutamate receptor expressed in a cloned baby hamster kidney cell line. *Brain Res.* **619**, 22–28.

Watkins, J. C., Krogsgaard-Larsen, P., and Honore, T. (1990) Structure–activity relationships in the development of excitatory amino acid receptor agonists and competitive antagonists. *Trends Pharmacol. Sci.* **11**, 25–33.

Watson, G. B., Monaghan, D. T., and Lanthorn, T. H. (1990) Selective activation of oscillatory currents by trans-ACPD in rat brain mRNA-injected *xenopus* oocytes and their blockade by NMDA. *Eur. J. Pharmacol.* **179**, 479–481.

Weiss, S. (1989) Two distinct quisqualate receptor systems are present on striatal neurons. *Brain Res.* **491**, 189–193.

Weiss, J. H., Christine, C. W., and Choi, D. W. (1989) Bicarbonate dependence of glutamate receptor activation by β-*N*-methylamino-L-alanine: channel recording and study with related compounds. *Neuron* **3**, 321–326.

Winder, D. G. and Conn P. J. (1992) Activation of metabotropic glutamate receptors in the hippocampus increases cyclic AMP accumulation. *J. Neurochem.* **59**, 375–378.

Winder, D. G., Smith, T., and Conn, P. J. (1993) Pharmacological differentiation of metabotropic glutamate receptors coupled to potentiation of cyclic adenosine monophosphate responses and phosphoinositide hydrolysis. *J. Pharmacol. Exp. Ther.* **266**, 518–525.

Wroblewska, B., Wroblewski, J. T., Saab, O. H., and Neale, J. H. (1993) *N*-Acetylaspartylglutamate inhibits forskolin-stimulated cyclic AMP levels via a metabotropic glutamate receptor in cultured cerebellar granule cells. *J. Neurochem.* **61**, 943–948.

CHAPTER 3

Second-Messenger Systems Coupled to Metabotropic Glutamate Receptors

P. Jeffrey Conn, Valerie Boss, and Dorothy S. Chung

1. Introduction

Glutamate and other excitatory amino acids (EAAs) have long been known to increase the levels of various second-messenger systems in different nervous system preparations. However, until recent years, these effects were generally held to be secondary to activation of glutamate-gated cation channels, and subsequent increases in neurotransmitter release or intracellular calcium concentrations. The first direct evidence for the existence of glutamate receptors directly coupled to second-messenger systems via GTP-binding proteins (G-proteins) came in the mid1980s with the discovery of glutamate receptors coupled to activation of phosphoinositide hydrolysis. Since that time, it has become clear that members of the metabotropic glutamate receptor (mGluR) family are coupled, either directly or indirectly, to a variety of second-messenger systems, including activation of phosphoinositide hydrolysis, regulation of adenylyl cyclase,

The Metabotropic Glutamate Receptors Eds.: P. J. Conn and J. Patel
© 1994 Humana Press Inc., Totowa, NJ

activation of phospholipase D, increased cyclic guanosine monophosphate (cGMP) accumulation, and arachidonic acid release.

2. Phosphoinositide Hydrolysis as a Transduction Mechanism for mGluRs

2.1. Activation of Phosphoinositide Hydrolysis Can Result in Formation of Multiple Second Messengers

Receptor-activated phosphoinositide hydrolysis serves as a major signal transduction mechanism in the brain and other organ systems (*see* Chuang, 1989; Rana and Hokin, 1990 for reviews). Interaction of an agonist with a phosphoinositide hydrolysis-linked receptor leads to activation of a phosphoinositide-specific phosphodiesterase (phospholipase C) that catalyzes the hydrolysis of membrane phosphoinositides. Phosphatidylinositol-4,5-*bis*-phosphate (PIP_2) is the primary substrate for the activated enzyme and both products of PIP_2 hydrolysis, inositol-1,4,5-trisphosphate (IP_3) and diacylglycerol (DAG), act as second messengers. IP_3 releases calcium from intracellular stores and thereby elevates cytosolic calcium. Calcium ions then interact with a variety of intracellular targets (e.g., protein kinases, phosphatases, ion channels, and so forth) to alter cell function. The second product formed by hydrolysis of phosphoinositides, DAG, lowers the calcium requirement of a phospholipid/calcium-dependent protein kinase termed protein kinase C (PKC) and thereby increases its activity. DAG can also activate certain calcium-independent forms of PKC.

Activation of phosphoinositide hydrolysis results in liberation of other intracellular messengers that alter cell function. For instance, arachidonic acid is released subsequent to the rise in DAG levels owing to the action of a DAG-specific phospholipase. Arachidonic acid serves as a substrate for cyclooxygenase and lipoxygenase, leading to production of prostaglandins and leukotrienes. These compounds have numerous cellular effects. Furthermore, the increased levels of arachidonic acid, coupled with the increased levels of intracellular calcium, may activate guanylyl cyclase and consequently increase formation of cyclic GMP. Other inositol phosphate products, such as inositol-1,3,4,5-tetrakisphosphate and inositol-1:2-cyclic-4,5-trisphosphate, are also formed, and some of these may have intracellular effects.

2.2. Discovery of a Novel Glutamate Receptor Coupled to Activation of Phosphoinositide Hydrolysis

The first indication that glutamate stimulates phosphoinositide hydrolysis came from studies in primary cultures of striatal neurons where it was shown that glutamate, *N*-methyl-D-aspartate (NMDA), kainate (KA), and quisqualate stimulate formation of IP_3 (Sladeczek et al., 1985). The most potent and efficacious agents in stimulating this response were quisqualate and glutamate. The response to these two agents was insensitive to blockade by a selective NMDA receptor antagonist, whereas the response to NMDA was partially blocked by the NMDA receptor antagonist, leading to the suggestion that activation of both quisqualate- and NMDA-preferring receptors can lead to activation of phosphoinositide hydrolysis. Subsequent early studies in *Xenopus* oocytes injected with rat brain messenger RNA (mRNA) provided clear evidence that the response to quisqualate is at least partially mediated by activation of a G-protein coupled receptor that is pharmacologically distinct from the NMDA and KA/α-amino-3-hydroxyl-5-methyl-isoxazole-4-propionic acid (AMPA) receptor subtypes (Sugiyama et al., 1987). In oocytes injected with rat brain mRNA, but not in control oocytes, glutamate activated phosphoinositide hydrolysis. This effect was mimicked by quisqualate, but not by NMDA or kainate. Although the response was induced by quisqualate, it was not mediated by the quisqualate-preferring AMPA receptor, since it was not blocked by an antagonists of the ionotropic glutamate receptors. In contrast, the effect of glutamate or quisqualate was inhibited by pertussis toxin, suggesting involvement of a G-protein-linked receptor. Since these early studies, EAA-stimulated phosphoinositide hydrolysis has been characterized in a number of other preparations, including cultured astrocytes (Nicoletti et al., 1990; Pearce et al., 1990), cultured neurons (Furuya et al., 1989; Patel et al. 1990), synaptoneurosomes (Recasens et al., 1988; Dudek and Bear, 1989), and slices from a variety of rat brain regions (Nicoletti et al., 1986b; Blackstone et al., 1989; Palmer et al., 1990; Desai and Conn, 1990).

It is likely that EAA-induced increases in phosphoinositide hydrolysis are mediated by direct coupling to phospholipase C via a GTP-binding protein. In most systems examined, EAA-stimulated phosphoinositide hydrolysis is only slightly decreased by omission of added calcium from the extracellular medium (Nicoletti et al., 1986b; Palmer et al., 1988; Ambrosini and Meldolesi, 1989; Doble and Perrier, 1989), addition of tetrodotoxin to the medium (Patel et al., 1991), or addition of antagonists at other known neurotransmitter receptors (Nicoletti et al., 1986a). These data suggest that this response is not dependent on calcium-dependent neurotransmitter release, or calcium influx and subsequent activation of phospholipase C. The phosphoinositide hydrolysis response to EAAs is partially blocked by pertussis toxin in a variety of preparations (Sugiyama et al., 1987; Nicoletti et al., 1988; Ambrosini and Meldolesi, 1989; Houamed et al., 1991; Masu et al., 1991). Furthermore, in permeablized astrocytes, glutamate-stimulated phosphoinositide hydrolysis is potentiated by guanosine 5'-O-(3-thiotriphosphate), a nonhydrolyzable GTP analog, and inhibited by 5'-O-(3-thiodiphosphate) (Robertson et al., 1990). Since activation of phospholipase C by calcium does not require involvement of a G-protein, this suggests that the response is mediated by activation of a G-protein-linked receptor rather than being secondary to calcium influx.

It should be noted that in most cases in which it has been tested, omission of added extracellular calcium did diminish EAA-stimulated phosphoinositide hydrolysis at least slightly. In at least two reports, reduction of extracellular calcium virtually abolished EAA-stimulated phosphoinositide hydrolysis (Alexander et al., 1990; Patel et al., 1991). Based on this, it has been proposed that intracellular entry of calcium does play some role in mGluR-mediated phosphoinositide hydrolysis (Guiramand et al., 1991; Patel et al., 1991). It has been established that increased intracellular calcium not only activates phospholipase C directly, but also increases GTP-dependent phospholipase C activity (McDonough et al., 1988). Thus, it is possible that calcium entering the cell from an extracellular source plays a synergistic role with the mGluRs in activating phospholipase C. Consistent with this, calcium markedly potentiates the effect of glutamate on phosphoinositide hydrolysis in permeablized astrocytes, but calcium alone has a smaller effect on phosphoinositide hydrolysis than does glutamate alone (Robertson et al., 1990).

2.3. Pharmacological Profile of Phosphoinositide Hydrolysis-Linked mGluRs

In most preparations, phosphoinositide hydrolysis-linked mGluRs are activated by a variety of EAAs that act at ionotropic glutamate receptors, including glutamate, quisqualate, *cis*-1-amino-1,3-cyclopentanedicarboxylic acid (*cis*-ACPD), and ibotenate (Nicoletti et al., 1986b; Schoepp and Johnson, 1988; Sugiyama et al., 1989; Weiss, 1989; Desai and Conn, 1990; Pearce et al., 1990). Although in some of these preparations, a small phosphoinositide hydrolysis response can be evoked secondary to activation of one of the ionotropic glutamate receptors (Sladeczek et al., 1985; Nicoletti and Canonico, 1989; Milani et al., 1990), in the majority of cases, the primary response is mediated by an mGluR since it is not mimicked by more selective ionotropic glutamate receptor agonists such as AMPA and NMDA (Nicoletti et al., 1986b; Schoepp and Johnson, 1988; Sugiyama et al., 1989; Weiss, 1989; Pearce et al., 1990), or blocked by a wide variety of ionotropic glutamate receptor antagonists (Nicoletti et al., 1986b; Schoepp and Johnson, 1988; Doble and Perrier, 1989; Sugiyama et al., 1989; Weiss, 1989; Desai and Conn, 1990; Godfrey and Taghavi, 1990; Pearce et al., 1990). Furthermore, two EAAs have have now been shown to be selective mGluR agonists that activate phosphoinositide hydrolysis at concentrations that have no effect on ionotropic glutamate receptors. These include *trans*-1-amino-1,3-cyclopentanedicarboxylic acid (*trans*-ACPD) (Palmer et al., 1989; Desai and Conn, 1990; Manzoni et al., 1991) or its active isomer 1S,3R-ACPD (Irving et al., 1990; Desai et al., 1992; but *also see* Manzoni et al., 1992a), and the (2S,3S,4S) isomer of α-(carboxycyclopropyl)glycine (L-CCG-I) (Nakagawa et al., 1990), both of which are rigid analogs of the extended conformation of glutamate.

At present, there are few antagonists available for the mGluRs (*see* Chapter 2). Phosphoserine, α-aminoadipate, 2-amino-4-phosphonobutanoate (AP4) (Nicoletti et al., 1986a; Schoepp and Johnson, 1988, 1989a), and L-aspartate-β-hydroxamate (Ormandy, 1992; Porter et al., 1992) have weak partial agonist or antagonist actions on phosphoinositide hydrolysis-linked mGluRs in rat brain slices. However, each of these compounds either has too much intrinsic activity or lacks the selectivity and/or potency needed to serve as a useful antagonist. A related compound that has been useful as an antagonist of some phosphoinositide hydrolysis-coupled mGluRs is 2-amino-3-phosphonopropionate (L-AP3) (Schoepp and Johnson, 1989a,b). L-AP3 is a noncompetitive, irreversible antagonist at phosphoinositide hydrolysis-linked mGluRs in hippocampal slices and lacks appreciable affinity for ionotropic glutamate receptors. L-AP3 also at least partially blocks mGluR-mediated increases in phosphoinositide hydrolysis in cerebellar granule cells (Irving et al., 1990; Aronica et al., 1993), hypothalamic slices (Sortino et al., 1991), striatal neurons (Manzoni et al., 1991), *Xenopus* oocytes injected with rat brain mRNA (Tanabe et al., 1992), and cells expressing mGluR1 (Tanabe et al., 1993). However, as discussed below, mGluRs coupled to activation of phosphoinositide hydrolysis in several other preparations are insensitive to L-AP3, suggesting that multiple phosphoinositide hydrolysis-linked mGluRs are differentially sensitive to this compound.

More recently, various phenylglycine derivatives have been shown to serve as antagonists or weak partial agonists of phosphoinositide hydrolysis-coupled mGluRs (Gerber and Gähwiler, 1992; Bashir et al., 1993; Birse et al., 1993; Eaton et al., 1993). For instance, RS-α-methyl-4-carboxyphenylglycine (MCPG) (Bashir et al., 1993) and (D,L)4-carboxy-3-hydroxyphenylglycine (CHPG) (Thomsen and Suzdak, 1993) inhibit 1S,3R-ACPD-induced phosphoinositide hydrolysis in cell lines expressing mGluR1a. Furthermore, MCPG competitively blocks 1S,3R-ACPD-stimulated phosphoinositide hydrolysis in rat cerebral cortical slices and inhibits a variety of mGluR-mediated electrophysiological responses in thalamic (Eaton et al., 1993) and hippocampal (Bashir et al., 1993) neurons. A related compound, (S)-4-carboxyphenylglycine (CPG), also competitively inhibits 1S,3R-ACPD-induced phosphoinositide hydrolysis in cerebral cortical slices (Eaton et al., 1993).

2.4. Activation of mGluRs Leads to Mobilization of Intracellular Calcium and Translocation of Protein Kinase C

Evidence from a number of systems suggests that, as with other phosphoinositide hydrolysis-linked receptors, activation of mGluRs leads to release of calcium from intracellular stores. Experiments with calcium indicators, such as fura-2, indicate that glutamate and other mGluR agonists induce calcium mobilization from cultured hippocampal neurons (Murphy and Miller, 1988, 1989; Furuya et al., 1989; Glaum et al., 1990b), cerebellar neurons (Bouchelouche et al., 1989; Irving et al., 1990), cortical synaptosomes (Adamson et al., 1990), and hippocampal or cortical (but not cerebellar) astrocytes (Glaum et al., 1990a). Measurement of the effect of EAAs on a calcium-induced chloride current in *Xenopus* oocytes injected with rat brain mRNA (Sugiyama et al., 1987, 1989; Verdoorn and Dingledine, 1988; Manzoni et al., 1990) or in which mGluR1a is expressed (Houamed et al., 1991; Masu et al., 1991) indicates that a similar response occurs in these cells. In most of the fura-2 experiments, glutamate and other mixed ionotropic/metabotropic agonists induce two distinct effects on intracellular calcium levels. These include an initial transient calcium spike or group of oscillations that are followed by a prolonged plateau phase. In general, selective ionotropic agonists only induce the second phase, and this is completely abolished by removal of calcium from the extracellular medium or by addition of ionotropic receptor antagonists. In contrast, the initial phase of the mixed ionotropic/metabotropic response is generally insensitive to removal of extracellular calcium, suggesting that it is mediated by mobilization of intracellular calcium stores. In addition, this phase is not mimicked by selective ionotropic agonists or blocked by ionotropic receptor antagonists, but is mimicked by the selective metabotropic glutamate receptor agonist 1S,3R-ACPD (Irving et al., 1990, 1992; Manzoni et al., 1990). Thus, the initial calcium spike is likely to be mediated by activation of mGluRs and mobilization of intracellular calcium. Evidence suggests that this mGluR-induced calcium mobilization is mediated by stimulation of release of calcium from the IP_3-sensitive intracellular pool, since it is diminished after depletion of IP_3-sensitive calcium pools with agonists of other phosphoinositide hydrolysis-linked receptors or thapsigargin, an inhibitor of ATP-dependent uptake of calcium into intracellular stores (Irving et al., 1992). Furthermore, the response to mGluR agonists shows cross-desensitization with injected IP_3 in *Xenopus* oocytes (Sugiyama et al., 1987).

Less is known about the effect of mGluR activation on formation of DAG, the other second messenger associated with activation of phosphoinositide hydrolysis. However, as discussed above, the effects of DAG are generally mediated by activation of PKC. When activated, PKC is translocated from a cytosolic fraction to a membrane fraction. Nicoletti et al. (1990) found that mGluR agonists induce translocation of PKC in cultured astrocytes, suggesting that PKC is activated with mGluR activation. Also, glutamate induces a transient rise in DAG levels and activation of PKC in cultured hippocampal neurons (Scholz and Palfrey, 1991). However, it is not yet known whether this effect of glutamate is mediated by activation of mGluRs or ionotropic glutamate receptors.

2.5. Evidence
for Multiple mGluRs
Coupled to Activation of Phosphoinositide Hydrolysis

As discussed above and in Chapter 1, of the cloned mGluRs, mGluR1a, mGluR1b, mGluR1c, mGluR5a, and mGluR5b are all capable of coupling to activation of phosphoinositide hydrolysis in *Xenopus* oocytes or cell lines, and it is likely that other clones and alternate splice variants exist that have yet to be characterized. Although receptors do not always couple to the same second-messenger systems in transfected cells as they do in cells in which they are normally expressed, it is likely that multiple phosphoinositide hydrolysis-linked mGluR subtypes are present in brain cells. There is increasing evidence from studies in primary cell cultures and brain slices that are consistent with this hypothesis. For instance, several studies indicate that mGluR-mediated increases in phosphoinositide hydrolysis are differentially sensitive to blockade by AP3 and AP4 in different preparations (Schoepp and Johnson, 1988; Murphy and Miller, 1989; Patel et al., 1990; Tanabe et al., 1992; Littman et al., 1993). Within a given region, there are clear differences in developmental regulation of phosphoinositide hydrolysis responses to different mGluR agonists (Schoepp and Hillman, 1990). Furthermore, there are clear differences in the agonist profiles of mGluRs coupled to activation of phosphoinositide hydrolysis in different preparations. For instance, in many preparations, phosphoinositide hydrolysis-linked mGluRs are selectively activated by 1S,3R-ACPD with 1R,3S-ACPD having no effect or serving as a weak partial agonist (Irving et al., 1990; Schoepp et al., 1991, 1992b). In contrast, Manzoni et al. (1992a) found that 1S,3R- and 1R,3S-ACPD have similar efficacy at stimulating phosphoinositide hydrolysis in primary cultures of cerebellar and striatal neurons, and that 1R,3S-ACPD was more potent than 1S,3R-ACPD at stimulating this response. More recently, Chung et al. (1993) found evidence for a phosphoinositide hydrolysis-coupled mGluR in rat cortical slices that is insensitive to activation by 1S,3R-ACPD, but is selectively activated by (RS)-4-bromohomoibotenate (BrHI). Although both 1S,3R-ACPD and BrHI activate phosphoinositide hydrolysis in this preparation, the response to these two agonists is completely additive. Furthermore, the phosphoinositide hydrolysis response to 1S,3R-ACPD is partially blocked by L-AP3, whereas the response to BrHI is L-AP3-insensitive. Earlier studies suggesting the existence of multiple mGluR subtypes coupled to activation of phosphoinositide hydrolysis have been discussed in detail by Schoepp et al. (1990). At present, it is not clear which of the cloned mGluR subtypes mediate EAA-induced phosphoinositide hydrolysis responses in different preparations. Although some speculation could be made based on available data, definitive conclusions must await development of more subtype-specific mGluR agonists and antagonists.

2.6. Physiological Roles
of Phosphoinositide Hydrolysis-Coupled mGluRs

At present, little is known about the specific roles of phospho-inositide hydrolysis-coupled mGluRs in regulating brain function. Regional analysis of *trans*-ACPD-stimulated phosphoinositide hydrolysis reveals that this response is more robust in hippocampal slices than in slices from other brain regions (Desai and Conn, 1990). Despite this, evidence suggests that the predominant phospho-inositide hydrolysis-linked mGluR in the hippocampus does not mediate any of a variety of electrophysiological responses to 1S,3R-ACPD. These include blockade of spike-frequency adaptation, reduction of the potassium currents I_M and I_{AHP}, spike broadening, and reduction of excitatory and inhibitory synaptic transmission. This conclusion is based on differences between mGluRs that mediate the phosphoinositide hydrolysis response and the various physiological responses in terms of both pharmacological profiles (Desai et al., 1992; Hu and Storm, 1992) and developmental regulation (Boss et al., 1992). Also, many of the physiological responses to mGluR activation are not blocked by PKC inhibitors (Gerber et al., 1992).

A physiological effect of mGluR activation in the hippocampus that may be mediated by activation of phosphoinositide hydrolysis-linked mGluRs is potentiation of NMDA-induced currents. 1S,3R-ACPD potentiates NMDA-induced inward currents (Aniksztejn et al., 1991, 1992) and depolarization (Harvey and Collingridge, 1993) in hippocampal pyramidal cells. Aniksztejn et al. (1992) reported that the effect of 1S,3R-ACPD on NMDA receptor currents is partially mimicked by application of PKC-activating phorbol esters, and blocked by intracellular injection of a mixture of the PKC inhibitors PKC(19-36) and sphingosine, suggesting a possible role for activation of phosphoinositide hydrolysis and subsequent activation of PKC. However, neither of the PKC inhibitors used is entirely specific for PKC relative to some other protein kinases (particularly calcium/calmodulin-dependent protein kinase II). Harvey and Collingridge (1993) found that two other protein kinase inhibitors, K-252b and staurosporine, were without effect on potentiation of NMDA-induced depolarization of hippocampal neurons by 1S,3R-ACPD. Furthermore, they also found that this response is not blocked by thapsigargin, a depletor of IP_3-sensitive calcium stores, arguing against a role for phosphoinositide hydrolysis-linked mGluRs. Thus, it is still unclear whether mGluR-induced potentiation of NMDA responses is mediated by activation of phosphoinositide hydrolysis or some other signaling pathway.

Phosphoinositide hydrolysis-linked mGluRs may also play a role in induction of long-term potentiation (LTP) (*see* Chapter 7 for detailed discussion). However, as with the question of NMDA receptor involvement in modulation of NMDA receptor currents, this issue is still under active investigation.

3. Coupling of mGluRs to Inhibition of Adenylyl Cyclase

A common mechanism of signal transduction employed by a variety of neurotransmitter receptors is attenuation of cyclic AMP (cAMP) accumulation, leading to functional antagonism of the effects of agents whose responses are mediated by increases in cAMP levels. Although a lowering of cAMP accumulation can be accomplished by activation of cyclic nucleotide phosphodiesterases, it is more often accomplished through receptor coupling to G_i, a G-protein that can inhibit adenylyl cyclase (for review, *see* Limbird, 1988).

3.1. Multiple mGluR Subtypes Are Coupled to Inhibition of Adenylyl Cyclase

Several of the cloned mGluRs are capable of coupling in a negative manner to adenylyl cyclase when expressed in cell lines (*see* Chapter 1). These include mGluR2 (Tanabe et al., 1992), mGluR3, mGluR4 (Tanabe et al., 1993), mGluR6 (Nakajima et al., 1993), and mGluR7 (Masu and Nakanishi, 1993). However, the finding that specific mGluRs can couple to inhibition of adenylyl cyclase in expression systems does not necessarily imply that this is the signal transduction mechanism employed by these receptors in cells in which they are natively expressed. Thus, it is important to determine the effect of mGluR activation on cAMP accumulation in brain slices and primary cell cultures. ACPD and other mGluR agonists have now been shown to inhibit forskolin-induced cAMP accumulation in hippocampal (Casabona et al., 1992; Schoepp et al., 1992a; Schoepp and Johnson, 1993), cerebral cortical (Cartmell et al., 1992), and hypothalamic (Casabona et al., 1992) slices as well as primary cultures of cerebellar granule cells (Wroblewska et al., 1993), striatal neurons (Manzoni et al., 1992b; Prezeau et al., 1992), and astrocytes (Baba et al., 1993). Several lines of evidence suggest that this effect is mediated by direct coupling of mGluRs to adenylyl cyclase via G_i. For instance, the effect is not likely to be mediated by activation of a cyclic nucleotide phosphodiesterase, since it is not occluded by phosphodiesterase inhibitors (Cartmell et al., 1992; Itano et al., 1992; Prezeau et al., 1992; Baba et al., 1993; Wroblewska et al., 1993). Furthermore, the effect can be measured in broken membrane preparations (Itano et al., 1992; Prezeau et al., 1992), suggesting that it does not involve production of an intracellular second messenger. Finally, mGluR-mediated inhibition of adenylyl cyclase is blocked by pertussis toxin (Itano et al., 1992; Manzoni et al., 1992b; Prezeau et al., 1992; Baba et al., 1993; Wroblewska et al., 1993), a toxin that irreversibly inactivates G_i and certain other G-proteins. Taken together, these data suggest that mGluRs that are negatively linked to adenylyl cyclase via G_i exist in brain slices and primary neuronal and glial cultures.

The finding that several of the cloned mGluRs can couple to inhibition of cAMP accumulation in expression systems suggests that multiple mGluR subtypes may be coupled to this signal transduction mechanism in mammalian brain preparations. Consistent with this, mGluR-mediated inhibition of cAMP accumulation has different pharmacological profiles in different preparations. For instance, in cultured striatal neurons, glutamate and quisqualate inhibit forskolin-stimulated cAMP accumulation with similar potencies, whereas L-AP3 and ibotenate are relatively inactive in these cells (Prezeau et al., 1992). In contrast, L-AP3, L-AP4, and ibotenate potently inhibit forskolin-stimulated cAMP accumulation in hippocampal slices, whereas quisqualate has relatively low potency in this preparation (Schoepp and Johnson, 1993; Genazzani et al., 1993). L-AP4 also potently inhibits cAMP accumulation in cultured cerebellar granule cells. Although it is not yet clear which of the cloned mGluRs mediate inhibition of cAMP accumulation in the different preparations, the pharmacological profiles of inhibition of cAMP accumulation by activation of various cloned receptors expressed in cell lines resemble those described in slice preparations and neuronal cultures (*see* Chapter 1). For instance, L-AP4 is a potent agonist of mGluRs 4, 6, and 7, whereas quisqualate has relatively little activity at these mGluRs (Thomsen et al., 1992; Nakajima et al., 1993; Tanabe et al., 1993). In contrast, quisqualate does activate mGluRs 2 and 3 (although with low potency), but L-AP4 has little or no activity at these mGluR subtypes expressed in cell lines (Tanabe et al., 1992, 1993).

A unique glutamate receptor has recently been characterized in dispersed hippocampal neurons that is clearly distinct from the cloned mGluRs and is negatively coupled to adenylyl cyclase (Itano et al., 1992). In this preparation, glutamate and NMDA inhibit forskolin-stimulated cAMP accumulation, whereas quisqualate and kainate are relatively inactive. This response is blocked by the selective NMDA receptor antagonist D-AP5. At first glance, these data would suggest that this response is mediated by the NMDA receptor. However, further analysis suggests that it may be mediated by a novel mGluR that has a similar pharmacological profile to that of NMDA receptors. Thus, although D-AP5 blocks this response, it is not blocked by the NMDA receptor channel blocker Mg^{2+}. Furthermore, the response is not blocked by removal of Ca^{2+} from the extracellular medium or incubation with tetrodotoxin. These data suggest that NMDA-induced inhibition of cAMP accumulation is not dependent on entry of Ca^{2+} through NMDA receptor channels, or increases in cell firing and subsequent neurotransmitter release. Perhaps most importantly, NMDA and glutamate inhibit forskolin-induced cAMP accumulation in membranes from the dispersed hippocampal cells, and this response is blocked by pertussis toxin. Thus, NMDA-induced reduction of cAMP accumulation may be mediated by a previously uncharacterized mGluR. Unfortunately, Itano et al. (1992) did not determine the effects of ACPD or other selective mGluR agonists to determine if these compounds interact with the NMDA-sensitive receptor.

3.2. Physiological Roles of mGluRs
That Are Negatively Coupled to Adenylyl Cyclase

At present, the physiological roles of mGluR-mediated reduction in cAMP accumulation are not known. However, cAMP plays an important role in a variety of functions in the brain, including modulation of ion channels, regulation of synaptic transmission, regulation of gene expression, and a variety of metabolic functions. Activation of mGluRs that are negatively coupled to adenylyl cyclase may be involved in inhibiting any of the responses to neurotransmitters that increase cAMP levels. Furthermore, previous studies suggest that many neurotransmitter receptors that are negatively coupled to adenylyl cyclase are also coupled via G-proteins to ion channels and are often present on nerve terminals where they reduce neurotransmitter release (Limbird, 1988). Thus, the mGluRs that serve as autoreceptors on glutamatergic terminals (*see* Chapter 6) or those that appear to be directly coupled to ion channels (*see* Chapter 5 and discussion below) could be the same as those that are coupled to inhibition of cAMP accumulation. Consistent with this, L-AP4 is an agonist at both adenylyl cyclase-coupled mGluRs and mGluRs coupled to inhibition of calcium conductances in hippocampal neurons (Trombley and Westbrook, 1992), and an agonist at mGluR autoreceptors at certain glutamatergic synapses (*see* Chapter 6).

4. mGluRs
Involved in Increasing cAMP Accumulation

Recent studies suggest that mGluR1 can couple to activation of adenylyl cyclase when expressed in Chinese hamster ovary (CHO) cells (Aramori and Nakanishi, 1992), whereas none of the other cloned mGluRs positively couple to adenylyl cyclase in expression systems. However, a problem associated with studying second-messenger systems coupled to receptors in such expression systems is that overexpression of the receptor can result in coupling of the receptor to G-proteins and second-messenger systems that are not normally coupled to that receptor *in vivo*. As mentioned above, mGluR1 is also coupled to activation of phosphoinositide hydrolysis in CHO cells as well as a variety of other expression systems (*see* Chapter 1). Furthermore, mGluR1 can couple to increased arachidonic acid release in CHO cells. Thus, it is not clear whether mGluR1 is coupled to activation of adenylyl cyclase *in vivo*.

4.1. Activation of mGluRs Potentiates cAMP Responses to Agonists of Receptors That Are Positively Coupled to Adenylyl Cyclase

At present, there is no compelling evidence to suggest the presence of an mGluR that is positively coupled to adenylyl cyclase in primary cell cultures or tissue slices from mammalian brain. However, in recent months, a novel mGluR has been characterized that markedly potentiates cAMP responses to agonists of other receptors that are positively coupled to adenylyl cyclase via the stimulatory G-protein G_s (Alexander et al., 1992; Winder and Conn, 1992, 1993; Winder et al., 1993). Thus, 1S,3R-ACPD and other mGluR agonists markedly potentiate cAMP responses to a variety of other agonists in rat brain slices, including vasoactive intestinal polypeptide (VIP), 2-chloroadenosine (2-CA), the β-adrenergic receptor agonist isoproterenol, and prostaglandin E2 (Winder and Conn, 1993; Winder et al., 1993) (Fig. 1). This is reminiscent of previously described effects of norepinephrine and histamine, which potentiate cAMP responses by activation of α_1-adrenergic and H1 histaminergic receptors, respectively (Magistretti and Schorderet, 1985; Pilc and Enna, 1986; Johnson and Minneman, 1987; Garbarg and Schwartz, 1988; Schaad et al., 1989).

Although mGluRs positively coupled to adenylyl cyclase via G_s have not been found in rat brain slices, it has long been known that EAAs can increase basal cAMP accumulation in rat hippocampal slices in the absence of other exogenously added agonists (Schmidt et al., 1977; Bruns et al., 1980). However, this response was found to be dependent on the presence of endogenous adenosine. Since mGluRs had not been discovered when these studies were performed,

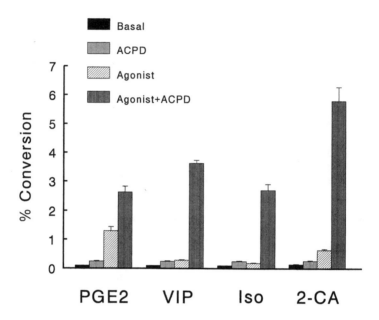

Fig. 1. Activation of mGluRs potentiates agonist-induced increases in cAMP accumulation. The effect of 100 µM 1S,3R-ACPD on the cAMP accumulation was measured in hippocampal slices in the presence or absence of agonists of receptors that are positively coupled to adenylyl cyclase via G_s. PGE2 = prostaglandin E_2, VIP = vasoactive intestinal peptide, Iso = isoproterenol, 2-CA = 2-chloroadenosine. Data are taken from Winder et al. (1993).

it was concluded that EAA-stimulated increases in cAMP accumulation were mediated by activation of ionotropic glutamate receptors, subsequent depolarization, and release of endogenous adenosine. However, recent studies indicate EAA-induced increases in basal cAMP accumulation are mediated by mGluR-induced potentiation of cAMP responses to adenosine that is already present in the extracellular space (Winder and Conn, 1992, 1993). Thus, extracellular adenosine concentrations in hippocampal slices are just below those needed to elicit a detectable cAMP response. Activation of mGluRs potentiates the cAMP response to these low levels of adenosine, resulting in a two- to fourfold increase in cAMP accumulation. It should be noted that estimates of extracellular concentrations of adenosine *in vivo* are similar to those in hippocampal slices (Park and Gidday, 1990), Furthermore, activation of ionotropic glutamate receptors is capable of evoking adenosine release (Hoehn and White, 1990a,b). Thus, under normal physiological conditions, the combined actions of glutamate on adenosine release and potentiation of cAMP responses to adenosine could lead to a large increase in cAMP accumulation. This would occur in the absence of activation of afferents from nonglutamatergic neurons that release neurotransmitters that activate G_s-coupled receptors. In addition, coactivation of glutamatergic afferents with afferents containing norepinephrine, VIP, or other neuromodulators could result in large increases in cAMP accumulation in postsynaptic cells that may not be achieved by activation of the neuromodulatory afferents alone.

4.2. Cellular Mechanisms of mGluR-Mediated Potentiation of cAMP Responses

At present, the mechanism by which mGluR activation potentiates cAMP responses is not known. However, previous studies have shown that PKC-activating phorbol esters potentiate cAMP responses in a manner similar to that seen with mGluR agonists (Garbarg and Schwartz, 1988; Donaldson et al., 1990). Thus, it has been suggested that EAA-induced potentiation of cAMP responses may be mediated by activation of phosphoinositide hydrolysis and subsequent activation of PKC (Donaldson et al., 1990; Alexander et al., 1992). Consistent with this, the diterpine forskolin inhibits both glutamate- and phorbol ester-induced potentiation of cAMP responses in guinea pig cortical slices (Donaldson et al., 1990). However, mGluR-mediated potentiation of cAMP responses in rat cortical slices is not blocked by the nonselective protein kinase inhibitor staurosporine at concentrations that are maximally effective at inhibiting PKC (Winder and Conn., 1992; unpublished observations). Furthermore, experiments using calcium indicators suggest that both 1S,3R-ACPD and the muscarinic agonist carbachol stimulate phosphoinositide hydrolysis in hippocampal pyramidal cells (Charpak et al., 1990; Knöpfel et al., 1990), but carbachol does not mimic the effects of 1S,3R-ACPD on cAMP accumulation in rat hippocampal slices (Winder and Conn, 1992). Finally, a detailed comparison of the pharmacological profile of mGluR-mediated phosphoinositide hydrolysis and potentiation of cAMP responses in rat cortical slices suggests that these responses are clearly mediated by distinct receptor subtypes (Winder et al., 1993). For instance, the rank order of agonist potencies at stimulating these two responses is clearly different, and L-serine-O-phosphate is a competitive antagonist at mGluRs coupled to potentiation of cAMP responses at concentrations that have little or no effect on mGluR-mediated increases in phosphoinositide hydrolysis. Taken together, these data suggest that 1S,3R-ACPD-induced potentiation of cAMP responses is not secondary to activation of the major phosphoinositide hydrolysis-linked mGluR in rat brain slices. However, these data do not completely rule out the possibility that this response is mediated by a phosphoinositide hydrolysis-linked mGluR that does not contribute significantly to the overall phosphoinositide hydrolysis response measured in hippocampal slices.

Another possible mechanism by which receptor activation could potentiate cAMP responses was recently proposed by Tang and Gilman (1991), who showed in cultured insect ovarian Sf9 cells that although the $\beta\gamma$ subunits of a heterotrimeric ($\alpha\beta\gamma$) G-protein inhibit type I adenylyl cyclase, these same subunits potentiate activation of type II adenylyl cyclase by the α subunit of G_s. Furthermore, Federman et al. (1992) showed that activation of α_2-adrenergic receptors increases cAMP levels in cells transfected with the type II adenylyl cyclase in a pertussis toxin-sensitive manner. Further studies in their system suggested that this is mediated by generation of free $\beta\gamma$ subunits, which potentiate α_s-mediated activation of type II adenylyl cyclase. Type II adenylyl cyclase is abundant in rat brain (Feinstein et al., 1991). Thus, in future experiments it will be interesting to determine whether the 1S,3R-ACPD-induced enhancement of cAMP responses to other neurotransmitters is mediated by a similar generation of $\beta\gamma$ subunits and activation of type II adenylyl cyclase.

4.3. Physiological Roles of mGluRs
Coupled to Potentiation of cAMP Responses

One of the most interesting aspects of the finding that mGluR activation potentiates cAMP responses to agonists of other receptors is its possible implications for associative synaptic plasticity. Associative forms of learning are thought to involve the induction of long-lasting changes in neuronal excitability or synaptic function by simultaneous activation of two independent inputs to a cell. Agonist-induced increases in intracellular cAMP concentrations are known to lead to lasting changes in synaptic responses in the hippocampus (Heginbotham and Dunwiddie, 1991; Slack and Pockett, 1991; Chavez-Noriega and Stevens, 1992; Dunwiddie et al., 1992; Haas and Gähwiler, 1992). It is conceivable that coactivation of glutamatergic afferents with afferents that release transmitters that activate G_s-coupled receptors could induce a large increase in cAMP accumulation and thereby elicit a change in synaptic function that would not normally be elicited by activation of either afferent alone. This could provide a novel mechanism for an associative form of synaptic plasticity that is induced by simultaneous activity of two independent afferents to a cell. Consistent with this hypothesis, Gereau and Conn (1994) recently reported that coactivation of mGluRs and β-adrenergic receptors induces a synergistic increase in cAMP accumulation in hippocampal slices and a lasting potentiation of evoked population spikes in hippocampal area CA1 (Fig. 2). This response can be mimicked by cell-permeable cAMP analogs or the adenylyl cyclase-activator forskolin (Slack and Pockett, 1991; Dunwiddie et al., 1992) and is blocked by the nonspecific protein kinase inhibitor staurosporine (Gereau and Conn, 1994). These results are consistent with the hypothesis that these effects are mediated by increases in cAMP levels and subsequent activation of a cAMP-dependent protein kinase.

The mechanisms involved in the enhancement of evoked population spikes by coactivation of mGluRs and β-adrenergic receptors are clearly distinct from the mechanisms involved in hippocampal

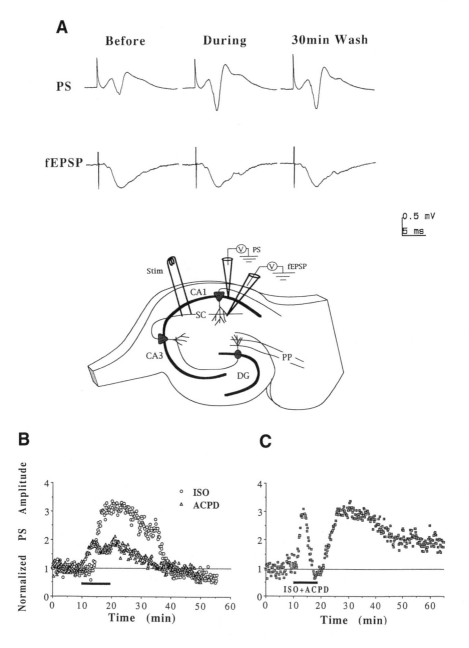

LTP in that the increased population spike amplitude is not accompanied by an increase in transmission at the Schaffer collateral-CA1 synapse (*see* Fig. 2). Furthermore, coapplication of ISO and 1S,3S-ACPD does not induce a lasting reduction of evoked IPSPs recorded in CA1 pyramidal cells, suggesting that the increase in population spike amplitude is not mediated by a persistent disinhibition. Instead, coactivation of β-adrenergic receptors and mGluRs induces a persistent depolarization of CA1 pyramidal cells that is accompanied by a slight increase in input resistance. Both of these effects are likely to contribute to the population spike enhancement. In addition, coapplication of these drugs leads to a long-lived reduction in the slow AHP that follows a burst of action potentials in these cells and blockade of spike frequency adaptation. These latter effects are not likely to contribute to the potentiation of evoked population spikes, but could play an important role in the overall physiological effects of coactivation of mGluRs and β-adrenergic receptors, and could markedly potentiate responses of CA1 pyramidal cells to a sustained barrage of excitatory synaptic activity.

Fig. 2 *(opposite page)*. Coactivation of mGluRs and β-adrenergic receptors synergistically enhances population spike amplitude in hippocampal area CA1. Population spikes (PS) and field EPSPs (fEPSP) were recorded simultaneously from stratum pyramidale and stratum radiatum, respectively, of area CA1. Panel (**A**) shows evoked population spikes and fEPSPs before, during, and 30 min after coapplication of ISO (100 nM) plus 1S,3S-ACPD (100 μM). This drug combination induced a lasting increase in population spike amplitude that was not accompanied by an increase in fEPSPs. The accompanying diagram shows the major excitatory connections in the hippocampus, and the placement of stimulating and recording electrodes. Stim = stimulating electrode. DG = Dentate Gyrus. PP = perforant path. Panel (**B**) shows the time-course of effects of 100 nM ISO or 100 μM 1S,3S-ACPD alone on population spike amplitude. As can be seen, neither ISO nor 1S,3R-ACPD induced a lasting potentiation of PS amplitude when added alone. Panel (**C**) shows the time-course of effects of coapplication of ISO and 1S,3S-ACPD on PS amplitude. Each point represents the amplitude of an individual population spike elicited at 0.1 Hz. Data are taken from Gereau and Conn (1994).

It is not yet clear whether cAMP plays a role in any of the acute physiological effects of application of 1S,3R-ACPD alone. 1S,3R-ACPD can elicit a number of responses in hippocampal pyramidal cells, including cell depolarization (or a slow inward current in voltage clamp) and blockade of spike-frequency adaptation. Goh and Ballyk (1993) reported that, in CA1 pyramidal cells, these responses were blocked by Rp-cAMPS, a cAMP analog that inhibits cAMP-dependent protein kinase, and suggested that they were mediated by 1S,3R-ACPD-induced increases in cAMP accumulation. However, Gerber et al. (1992) recently reported that another protein kinase inhibitor, staurosporine, does not block these responses in CA3 pyramidal cells at concentrations that completely block electrophysiological responses to cAMP in the same preparation. Furthermore, L-AP3 completely blocks 1S,3R-ACPD-induced increases in cAMP accumulation in hippocampal slices (Winder and Conn, 1992), but has no effect on these electrophysiological responses to ACPD (Stratton et al., 1990; Desai et al., 1992; Hu and Storm, 1992). Thus, these responses are not likely to be mediated by 1S,3R-ACPD-induced increases in cAMP accumulation.

5. Activation of mGluRs
Increases Cyclic GMP Accumulation

Early studies with glutamate and a variety of ionotropic glutamate receptor agonists demonstrated that EAAs can increase cyclic guanosine monophosphate (cGMP) levels in cerebellum *in vivo* and in cerebellar slices (Schmidt et al., 1977; Garthwaite and Balazs, 1978; Briley et al., 1979; Foster and Roberts, 1980, 1981). The majority of this effect is blocked by ionotropic glutamate receptor antagonists and is dependent on extracellular calcium. Evidence suggests that the cGMP response to ionotropic glutamate receptor activation is secondary to calcium influx through voltage-dependent and/or NMDA receptor channels, calcium-dependent activation of nitric oxide synthase (NOS), and formation of nitric oxide (NO). NO can then serve as a messenger molecule that directly activates guanylyl cyclase (Bredt and Snyder, 1989; Garthwaite et al., 1989a,b; Wood et al., 1990). More recently, Okada (1992) reported that the selective mGluR agonist *trans*-ACPD increases cGMP levels in cerebellar slices, and this effect is not blocked by selective ionotropic glutamate receptor antagonists. As with the cGMP response to ionotropic glutamate receptor activation, ACPD-stimulated cGMP accumulation can be blocked by inhibitors of NOS, suggesting the involvement of NO. One possible source of the calcium needed for activation of NOS is activation of phosphoinositide hydrolysis and IP_3-mediated release of intracellular calcium. Consistent with this, the response to ACPD is not dependent on extracellular calcium and is blocked by two compounds that partially block ACPD-stimulated increases in phosphoinositide hydrolysis (AP3 and L-AP4).

At present, the exact physiological roles of mGluR-mediated increases in cGMP accumulation in the cerebellum are not entirely clear. However, one intriguing possibility is that this response is involved in induction of long-term depression (LTD) of transmission at the parallel fiber-Purkinje cell synapse. As discussed in Chapters 7 and 8, evidence suggests that coactivation of AMPA receptors and mGluRs is required for induction of cerebellar LTD. Furthermore, several studies suggest that induction of LTD requires activation of NOS and subsequent NO-induced increases in cGMP accumulation. Thus, it is possible that mGluR-mediated increases in cGMP formation contribute to this response. However, it is important to remember that AMPA receptor activation also increases cGMP levels in the cerebellum, and the cGMP needed for induction of LTD could be formed as a result of activation of AMPA receptors, mGluRs, or both.

A possible role for mGluR-mediated increases in cGMP accumulation in the rat nucleus tractus solitarius (NTS) was recently proposed by Glaum and Miller (1993b). As discussed in Chapters 6 and 8, mGluR activation has a number of effects on NTS neurons. These include induction of a small inward current, reduction of GABA receptor currents evoked by muscimol (I_{musc}), reduction of evoked excitatory postsynaptic currents (EPSCs), and enhancement of AMPA-evoked currents (I_{AMPA}). Several activators of guanylyl cyclase mimic the effect of mGluR agonists on EPSCs, I_{musc}, and I_{AMPA} without inducing the inward current (Glaum and Miller, 1993a).

However, these responses were not altered by manipulations that increase or decrease NO levels, suggesting that they are not mediated by NO-dependent activation of guanylyl cyclase. Another intercellular messenger that has been implicated in activating guanylyl cyclase is carbon monoxide (CO) (Verma et al., 1993). Interestingly, Glaum and Miller (1993b) found that the cGMP-sensitive effects of mGluR agonists in the NTS can be blocked by a selective inhibitor of heme oxygenase, an enzyme involved in CO synthesis. In contrast, this heme oxygenase inhibitor did not block the effects of a cell-permeable cGMP analog. These data are consistent with the hypothesis that mGluR activation regulates certain aspects of cell excitability and synaptic transmission in the NTS by a mechanism that depends on CO-induced activation of guanylyl cyclase.

6. Coactivation of mGluRs and AMPA Receptors Increases Arachidonic Acid Release

Activation of phospholipase A2 (PLA2) results in hydrolysis of membrane phospholipids (especially phosphatidylcholine) at the sn2 position to yield a lyso-phospholipid and a free fatty acid. Most commonly, the fatty acid that occupies this position is the 20-carbon fatty acid arachidonic acid (AA). AA can serve as an intracellular as well as an intercellular messenger and, consequently, may have a variety of important effects on cell function. Some of these effects, such as the activation of PKC or other enzymes, may be mediated by AA itself. However, most of the biological actions of AA release are likely to be mediated by specific AA metabolites, such as the prostaglandins, leukotrienes, or hydroperoxyeicosatetranoic acids, that leave the cell of synthesis and act on cell-surface receptors of cells within the immediate vicinity of the cell from which they were released.

The nonselective AMPA/mGluR agonist quisqualate induces a dramatic PLA2-dependent increase in arachidonic acid release in striatal neurons (Dumuis et al., 1990, 1993). The AMPA receptor antagonist CNQX completely blocks the response to quisqualate, suggesting that this response is dependent on activation of the AMPA subtype of ionotropic glutamate receptor. However, AMPA is not capable of increasing AA release, suggesting that AMPA receptor activation is not sufficient for inducing this response. Likewise, the selective mGluR agonist ACPD elicits only a slight increase in AA release when added alone. However, coapplication of AMPA and ACPD elicits a large increase in AA release that is comparable to the effect elicited by quisqualate. Thus, associative activation of AMPA receptors and mGluRs results in an increase in AA release that cannot be elicited by activation of either of these receptors alone. Evidence suggests that the increase in AA release induced by associative mGluR/AMPA receptor activation depends on the activity of a Na^+/Ca^{2+} exchanger (Dumuis et al., 1993).

The roles of AA and its metabolites in the central nervous system are still relatively unexplored. However, AA has been proposed as a possible retrograde messenger involved in induction of LTP (Madison et al., 1991). Since several studies implicate mGluRs as playing a role in LTP (*see* Chapter 7), this could be an important role for AA released by coactivation of AMPA receptors and mGluRs. Consistent with this, Herrero et al. (1992) recently found that application of AA to cortical synaptosomes in the presence of ACPD results in a dramatic increase in glutamate release. Furthermore, Collins and Davies (1993) found that brief (5 min) application of AA and ACPD to hippocampal slices results in a long-lasting (>1 h) increase in transmission at the Schaffer collateral-CA1 synapse that is similar to that seen with tetanus-induced LTP.

7. A Novel Metabotropic Receptor for L-Cysteine Sulfinic Acid is Coupled to Activation of Phospholipase D

7.1. Activation of Phospholipase D Represents a Novel Mechanism for Signal Transduction

Increasing evidence suggests that a number of neurotransmitter and hormone receptors are coupled to activation of phospholipase D (PLD) and subsequent hydrolysis of membrane phospholipids, primarily phosphatidylcholine (PC). Activation of PLD results in generation of phosphatidic acid (PA) and free choline, and represents a novel mechanism employed by neurotransmitter receptors for signal transduction (for reviews, *see* Löffelholz, 1989; Billah and Anthes, 1990; Billah, 1993). PA, a product of PC metabolism, can have a number of cellular effects and has been proposed as a potential second-messenger compound. In addition, PA is converted by phosphatidate phosphohydrolase to DAG, which activates PKC and plays a key role in signal transduction (Fig. 3). Activation of various neurotransmitter receptors has been shown to increase DAG, but most such receptors identified to date are coupled to phospholipase C and activation of phosphoinositide hydrolysis. Activation of receptors coupled to the PLD-catalyzed metabolism of PC could potentially induce much higher levels of DAG than those associated with PLC-coupled phosphoinositide hydrolysis, since PC is the major phospholipid component of the cell membrane. Furthermore, DAG derived from different phospholipid substrates has different acyl side chains. Thus, it is possible that the DAG produced by the PLD-catalyzed breakdown of PC could preferentially activate different isoforms of PKC than that from the PLC-catalyzed metabolism of phosphoinositides.

7.2. A Novel Metabotropic EAA Receptor Is Coupled to Activation of PLD

Recent studies suggest that a novel metabotropic EAA receptor exists in hippocampal slices that is coupled to activation of PLD (Boss and Conn, 1992; Holler et al., 1993). The pharmacological profile of this response suggests that it is mediated by a receptor belonging to the mGluR family. For instance, this receptor is activated by 1S,3R-ACPD and 1S,3S-ACPD, but not by the 1R,3S-ACPD. Furthermore, the PLD response can be elicited by a variety of other mGluR agonists, including quisqualate, ibotenate, and L-CCG-I, but not by selective ionotropic glutamate receptor agonists, such as NMDA or kainate (Boss and Conn, 1992; Boss et al., 1992). Finally, the response is not blocked by selective ionotropic glutamate receptor antagonists (Boss and Conn, 1992). Interestingly, the putative mGluR antagonist L-AP3 is a full agonist at stimulating PLD activity.

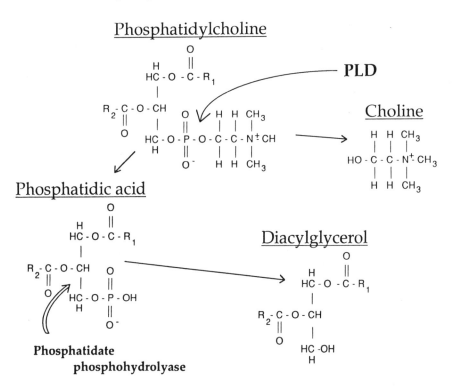

Fig. 3. Hydrolysis of membrane phosphatidylcholine by phospholipase D (PLD) results in the generation of a variety of metabolites, including choline, phosphatidic acid, and diacylglycerol.

Although the pharmacological profile of the PLD-coupled receptor is consistent with its being an mGluR, glutamate is not capable of increasing PLD activity in adult hippocampal slices (Boss and Conn, 1992; Boss et al., 1992), suggesting that glutamate may not be the endogenous agonist of this receptor. In contrast, another endogenous EAA, L-cysteine-sulfinic acid, has high efficacy at the PLD-coupled receptor and elicits a response similar in magnitude to that of 1S,3R-ACPD (Fig. 4) (Boss et al., 1994). L-CSA is a sulfur-containing EAA that fulfills virtually all of the criteria for serving as a neurotransmitter in mammalian brain (for reviews, *see* Recasens et al., 1982; Baba, 1987; Griffiths, 1990). The necessary enzymes for

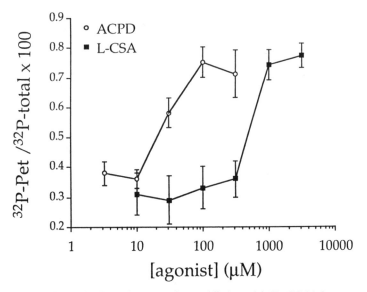

Fig. 4. 1S,3R ACPD and L-cysteine sulfinic acid (L-CSA) increase PLD activity in a concentration-dependent manner. PLD activity was measured as described by Boss and Conn (1992).

L-CSA synthesis and degradation are present in nerve endings, and L-CSA is released in a calcium-dependent manner on depolarization. Furthermore, L-CSA is a substrate for a high-affinity Na^+-dependent uptake system. L-CSA is structurally similar to L-glutamate and L-aspartate, and has long been known to have excitotoxic effects similar to these EAA neurotransmitters. In addition, L-CSA induces a variety of physiological and biochemical responses. However, until discovery of the PLD-coupled receptor, all effects of L-CSA that had been fully characterized were mediated by cross-reactivity with ionotropic glutamate receptors (Pullan et al., 1987; Griffiths, 1990; Curras and Dingledine, 1992; Porter and Roberts, 1993), and no receptors that were selectively activated by L-CSA had been characterized. Thus, activation of the PLD-coupled receptor may be the first known action of L-CSA that is independent of those mediated by ionotropic glutamate receptors.

The finding that glutamate does not significantly activate the PLD-coupled metabotropic receptor in brain slices from adults does not necessarily imply that glutamate is not an endogenous agonist of this receptor. Brain slices have a high capacity for glutamate uptake, and it is possible that this prevents exogenously added glutamate from accumulating to concentrations needed for receptor activation in the synapse. Indeed, glutamate is also a relatively weak agonist at stimulating phosphoinositide hydrolysis in brain slices, but has high efficacy at mGluRs expressed in cell lines or in primary cell cultures, where glutamate uptake is less likely to be a major factor (*see* Conn and Desai, 1991 for review). Consistent with this, glutamate does increase PLD activity in slices from rats at early stages of development when glutamate uptake systems are not well developed (Boss et al., 1992; Holler et al., 1993). However, this reflects a general developmental regulation of the response to all agonists, and L-CSA is still the most efficacious agonist at stimulating PLD activity in slices from immature rats (Boss et al., 1993). Also, it is important to note that, like glutamate, L-CSA is an excellent substrate for high-affinity uptake systems in rat brain slices. In light of the large body of literature implicating L-CSA as a neurotransmitter, these data suggest that L-CSA may be more likely than L-glutamate to be the endogenous agonist of the PLD-coupled metabotropic EAA receptor. A common conception of EAA neurotransmitter function is that all EAAs act on members of an EAA receptor family that is comprised of various ionotropic glutamate receptors and mGluRs, all of which use L-glutamate as the primary endogenous agonist. If L-CSA, rather that L-glutamate, is eventually found to be the endogenous agonist of the PLD-coupled receptor, this will suggest that specific receptor families may exist for different endogenous EAAs. Thus, the actions of various EAAs could be more diverse than was previously appreciated. In future studies, it will be important to develop cell-culture systems that do not contain significant glutamate and L-CSA uptake systems to investigate this possibility further.

8. Summary

Studies with cloned mGluRs in expression systems as well as mGluRs in *in situ* preparations suggest that mGluRs are coupled to a variety of signal transduction systems. At present, little information is available regarding the specific physiological roles of mGluRs coupled to each of the second-messenger pathways. Furthermore, the specific mGluR subtypes that mediate second-messenger responses *in situ* are not yet known. However, as subtype-specific agonists and antagonists for the mGluRs become available, this will allow determination of the second-messenger systems employed by specific mGluR subtypes, as well as the physiological roles of these mGluRs and their corresponding second messengers in regulating CNS function.

Acknowledgments

Work in the authors' laboratory is supported by NIH grants NS-28405 and NS-31373, a grant from the Council for Tobacco Research (PJC), and an NIH NRSA postdoctoral fellowship (VB).

References

Adamson, P., Hajimohammadreza, I., Brammer, M. J., Campbell, I. C., and Meldrum, B. S. (1990) Presynaptic glutamate/quisqualate receptors: effects on synaptosomal free calcium concentrations. *J. Neurochem.* **55,** 1850–1854.

Alexander, S. P. H., Hill, S. J., and Kendall, D. A. (1990) Excitatory amino acid-induced formation of inositol phosphates in guinea-pig cerebral cortical slices: involvement of ionotropic or metabotropic receptors? *J. Neurochem.* **55,** 1439–1441.

Alexander, S. P. H., Curtis, A. R., Hill, S. J., and Kendall, D. A. (1992) Activation of a metabotropic excitatory amino acid receptor potentiates A_{2b} adenosine receptor-stimulated cyclic AMP accumulation. *Neurosci. Lett.* **146,** 231–233.

Ambrosini, A. and Meldolesi, J. (1989) Muscarinic and quisqualate receptor-induced phosphoinositide hydrolysis in primary cultures of striatal and hippocampal neurons. Evidence for differential mechanisms of activation. *J. Neurochem.* **53,** 825–833.

Aniksztejn, L., Bregestovski, P., and Ben-Ari, Y. (1991) Selective activation of quisqualate metabotropic receptor potentiates NMDA but not AMPA responses. *Eur. J. Pharmacol.* **205,** 327–328.

Aniksztejn, L., Otani, S., and Ben-Ari, Y. (1992) Quisqualate metabotropic receptors modulate NMDA currents and facilitate induction of long-term potentiation through protein kinase C. *Eur. J. Neurosci.* **4,** 500–505.

Aramori, I. and Nakanishi, S. (1992) Signal transduction and pharmacological characteristics of a metabotropic glutamate receptor, mGluR1, in transfected CHO cells. *Neuron* **8,** 757–765.

Aronica, E., Condorelli, D. F., Nicoletti, F., Albani, P. D., Amico, C., and Balazs, R. (1993) Metabotropic glutamate receptors in cultured cerebellar granule cells: developmental profile. *J. Neurochem.* **60,** 559–565.

Baba, A. (1987) Neurochemical characterization of cysteine sulfinic acid, an excitatory amino acid, in hippocampus. *Japan. J. Pharmacol.* **43,** 1–7.

Baba, A., Saga, H., and Hashimoto, H. (1993) Inhibitory glutamate response on cyclic AMP formation in cultured astrocytes. *Neurosci. Lett.* **149,** 182–184.

Bashir, Z. I., Bortolotto, Z. A., Davies, C. H., Berretta, N., Irving, A. J., Seal, A. J., Henley, J. M., Jane, D. E., Watkins, J. C., and Collingridge, G. L. (1993) Induction of LTP in the hippocampus needs synaptic activation of glutamate metabotropic receptors. *Nature* **363,** 347–350.

Billah, M. M. (1993) Phospholipase D and cell signalling. *Curr. Opinion Immunol.* **5,** 114–123.

Billah, M. M. and Anthes, J. C. (1990) The regulation and cellular function of phosphatidylcholine hydrolysis. *Biochem. J.* **269,** 281–291.

Birse, E. F., Eaton, S. A., Jane, D. E., St. Jones, P. L., Porter, R. H. P., Pook, P. C., Sunter, D. C., Udvarhelyi, M., Wharton, B., Roberts, P. J., Salt, T. E., and Watkins, J. C. (1993) Phenylglcine derivatives as new pharmacological tools for investigating the role of metabotropic glutamate receptors in the central nervous system. *Neuroscience* **3,** 481–488.

Blackstone, C. D., Supattapone, S., and Snyder, S. H. (1989) Inositol phospholipid-linked glutamate receptors mediate cerebellar parallel-fiber-purkinje-cell synaptic transmission. *Proc. Natl. Acad. Sci. USA* **86,** 4316–4320.

Boss, V. and Conn, P. J. (1992) Metabotropic excitatory amino acid receptor activation stimulates phospholipase D in hippocampal slices. *J. Neurochem.* **59,** 2340–2343.

Boss, V., Nutt, K. M., and Conn, P. J. (1993) L-Cysteine sulfinic acid as an endogenous agonist of a novel metabotropic receptor coupled to stimulation of phospholipase D activity. *Mol. Pharmacol.* (in press).

Boss V., Desai M. A., Smith, T. S., and Conn, P. J. (1992) Trans-ACPD-induced phosphoinositide hydrolysis and modulation of hippocampal pyramidal cell excitability do not undergo parallel developmental regulation. *Brain Res.* **594,** 181–188.

Bouchelouche, P., Belhage, B., Frandsen, A., Drejer, J., and Schousboe, A. (1989) Glutamate receptor activation in cultured cerebellar granule cells increases cytosolic free Ca^{2+} and activation of Ca^{2+} influx. *Exp. Brain Res.* **76,** 281–291.

Bredt, D. S. and Snyder, S. H. (1989) Nitric oxide mediates glutamate-linked enhancement of cGMP levels in the cerebellum. *Proc. Natl. Acad. Sci. USA* **86,** 9030–9033.

Briley, P. A., Kouyoumdjian, J. C., Maidamous, M., and Gonnard, P. (1979) Effect of L-glutamate and kainate on rat cerebellar cGMP levels in vivo. *Eur. J. Pharmacol.* **54,** 181–184.

Bruns, R. F., Pons, F., and Daly, J. W. (1980) Glutamate- and veratridine-elicited accumulations of cyclic AMP in brain slices: a role for factors which potentiate adenosine-responsive systems. *Brain Res.* **189,** 550–555.

Cartmell, J., Kemp, J. A., Alexander, S. P. H., Hill, S. J., and Kendall, D. A. (1992) Inhibition of forskolin-stimulated cyclic AMP formation by 1-aminocyclo-pentane-*trans*-1,3-dicarboxylate in guinea-pig cerebral cortical slices. *J. Neurochem.* **58,** 1964–1966.

Casabona, G., Genazzani, A. A., Di Stefano, M., Sortino, M. A., and Nicoletti, F. (1992) Developmental changes in the modulation of cyclic AMP formation by the metabotropic glutamate receptor agonist 1S,3R-aminocyclopentane- 1,3-dicar-boxlic acid in brain slices. *J. Neurochem.* **59,** 1161–1163.

Charpak, S., Gähwiler, B. H., Do, K. Q., and Knöpfel, T. (1990) Potassium conductances in hippocampal neurons blocked by excitatory amino-acid transmitters. *Nature* **347,** 765–767.

Chavez-Noriega, L. E. and Stevens, C. F. (1992) Modulation of synaptic efficacy in field CA1 of the rat hippocampus by forskolin. *Brain Res.* **574,** 85–92.

Chuang, D. (1989) Neurotransmitter receptors and phosphoinositide turnover. *Ann. Rev. Pharmacol. Toxicol.* **29,** 71–110.

Chung, D. S., Winder, D. G., and Conn, P. J. (1993) 4-Bromohomoibotenic acid selectively activates an ACPD-insensitive metabotropic glutamate receptor coupled to phosphoinositide hydrolysis in rat cortical slices. *J. Neurochem.* (in press)

Collins, D. R. and Davies, S. N. (1993) Co-administration of (1S,3R)-1-amino-cyclopentane-1,3-dicarboxylic acid and arachidonic acid potentiates synaptic transmission in rat hippocampal slices. *Eur. J. Pharmacol.* **240,** 325,326.

Conn, P. J. and Desai, M. A. (1991) Pharmacology and physiology of metabotropic glutamate receptors in mammalian central nervous system. *Drug Dev. Res.* **24,** 207–229.

Curras, M. C. and Dingledine, R. (1992) Selectivity of amino acid transmitters acting at *N*-methyl-D-aspartate and amino-3-hydroxy-5-methyl-4-isoxazolproprionate receptors. *Mol. Pharmacol.* **41,** 520–526.

Desai, M. A. and Conn, P. J. (1990) Selective activation of phosphoinositide hydroly-sis by a rigid analogue of glutamate. *Neurosci. Lett.* **109,** 157–162.

Desai, M. A., Smith, T. S., and Conn, P. J. (1992) Multiple metabotropic glutamate receptors regulate hippocampal function. *Synapse* **12,** 206–213.

Doble, A. and Perrier, M. L. (1989) Pharmacology of excitatory amino acid receptors coupled to inositol phosphate metabolism in neonatal rat striatum. *Neurochem. Int.* **15,** 1–8.

Donaldson, J., Kendall, D. A., and Hill, S. J. (1990) Discriminatory effects of forskolin and EGTA on the indirect cyclic AMP responses to histamine, noradrenaline, 5-hydroxytryptamine, and glutamate in guinea-pig cerebral cortical slices. *J. Neurochem.* **54,** 1484–1491.

Dudek, S. M. and Bear, M. F. (1989) A biochemical correlate of the critical period for synaptic modification in the visual cortex. *Science* **246,** 673–675.

Dumuis, A., Pin, J. P., Oomagari, K., Sebben, M., and Bockaert, J. (1990) Arachidonic acid released from striatal neurons by joint stimulation of ionotropic and metabotropic quisqualate receptors. *Nature* **347,** 181–183.

Dumuis, A., Sebben, M., Fagni, L., Prezeau, L., Manzoni, O., Cragoe, E. J., Jr., and Bockaert, J. (1993) Stimulation by glutamate receptors of arachidonic acid release depends on the Na⁺/Ca²⁺ exchanger in neuronal cells. *Mol. Pharmacol.* **43,** 976–981.

Dunwiddie, T. V., Taylor, M., Heginbotham, L. R., and Proctor, W. R. (1992) Long-term increases in excitability in the CA1 region of rat hippocampus induced by β-adrenergic stimulation: possible mediation by cAMP. *J. Neurosci.* **12,** 506–517.

Eaton, S. A., Jane, D. E., Jones, P. L. S. J., Porter, R. H. P., Pook, P. C.-K., Sunter, D. C., Udvarhelyi, P. M., Roberts, P. J., Salt, T. E., and Watkins, J. C. (1993) Competitive antagonism at metabotropic glutamate receptors by *(S)*-4-carboxyphenylglycine and *(RS)*-α methyl-4-carboxyphenylglycine. *Eur. J. Pharmacol.* **244,** 195–197.

Federman, A. D., Conklin, B. R., Schrader, K. A., Reed, R. R., and Bourne, H. R. (1992) Hormonal stimulation of adenylyl cyclase through Gi-protein βγ subunits. *Nature* **356,** 159–161.

Feinstein, P. G., Schrader, K. A., Bakalyar, H. A., Tang, W., Krupinski, J., Gilman, A. G., and Reed, R. R. (1991) Molecular cloning and characterization of a Ca²⁺/calmodulin-insensitive adenylyl cyclase from rat brain. *Proc. Natl. Acad. Sci. USA* **88,** 10,173–10,177.

Foster, G. A. and Roberts, P. J. (1980) Pharmacology of excitatory amino acid receptors mediating the stimulation of rat cerebellar cyclic GMP levels *in vitro. Life Sci.* **27,** 215–221.

Foster, G. A. and Roberts, P. J. (1981) Stimulation of rat cerebellar guanosine 3',5'-cyclic monophosphate (cyclic GMP) levels: effects of amino acid antagonists. *Br. J. Pharmacol.* **74,** 723–729.

Furuya, S., Ohmori, H., Shigemoto, T., and Sugiyama, H. (1989) Intracellular calcium mobilization triggered by a glutamate receptor in rat cultured hippocampal cells. *J. Physiol.* **414,** 539–548.

Garbarg, M. and Schwartz, J. (1988) Synergism between histamine H1- and H2-receptors in the cAMP response in guinea pig brain slices: effects of phorbol esters and calcium. *Mol. Pharmacol.* **33,** 36–43.

Garthwaite, J. and Balazs, R. (1978) Supersensitivity of the cyclic GMP response to glutamate during cerebellar maturation. *Nature* **275,** 328,329.

Garthwaite, J., Garthwaite, G., Palmer, R. M. J., and Moncada, S. (1989a) NMDA receptor activation induces nitric oxide synthesis from arginine in rat brain slices. *Eur. J. Pharmacol.* **172,** 413–416.

Garthwaite, J., Southam, E., and Anderton, M. (1989b) A kainate receptor linked to nitric oxide synthesis from arginine. *J. Neurochem.* **53,** 1952–1954.

Genazzani, A. A., Casabona, G., L'Episcopo, M. R., Condorelli, D. F., Dell'Albani, P., Shinozaki, H., and Nicoletti, F. (1993) Characterization of metabotropic glutamate receptors negatively linked to adenylyl cyclase in brain slices. *Brain Res.* **622,** 132–138.

Gerber, U. and Gähwiler, B. H. (1992) 4C3HPG (RS-4-carboxy-3-hydroxyphenylglcine), a weak agonist at metabotropic glutamate receptors, occludes the action of *trans*-ACPD in hippocampus. *Eur. J. Pharmacol.* **221,** 401,402.

Gerber, U., Sim, J. A., and Gähwiler, B. H. (1992) Reduction of potassium conductances mediated by metabotropic glutamate receptors in rat CA3 pyramidal cells does not require protein kinase A. *Eur. J. Neurosci.* **4**, 792–797.

Gereau, R. W. and Conn, P. J. (1994) A cyclic AMP-dependent form of associative synaptic plasticity induced by coactivation of β-adrenergic receptors and metabotropic glutamate receptors in rat hippocampus. *J. Neurosci.* **14**, 3310–3318.

Glaum, S. R. and Miller, R. J. (1993a) Zinc protoporphyrin-1X blocks the effects of metabotropic glutamate receptor activation in the rat nucleus tractus solitarius. *Mol. Pharmacol.* **43**, 965–969.

Glaum, S. R. and Miller, R. J. (1993b) Activation of metabotropic glutamate receptor produces reciprocal regulation of ionotropic glutamate and GABA responses in the nucleus of the tractus solitarius of the rat. *J. Neurosci.* **13(4)**, 1636–1641.

Glaum, S. R., Holzwarth, J. A., and Miller, R. J. (1990a) Glutamate receptors activate Ca^{2+} mobilization and Ca^{2+} influx into astrocytes. *Proc. Natl. Acad. Sci. USA* **87**, 3454–3458.

Glaum, S. R., Scholz, W. K., and Miller, R. J. (1990b) Acute and long-term glutamate-mediated regulation of $[Ca^{++}]i$ in rat hippocampal pyramidal neurons in vitro. *J. Pharmacol. Exp. Ther.* **253**, 1293–1302.

Godfrey, P. P. and Taghavi, Z. (1990) The effect of non-NMDA antagonists and phorbol esters on excitatory amino acid-stimulated inositol phosphate formation in rat cerebral cortex. *Neurochem. Int.* **16**, 65–72.

Goh, J. W. and Ballyk, B. A. (1993) A cAMP-linked metabotropic glutamate receptor in hippocampus. *NeuroReport* **4**, 454–456.

Griffiths, R. (1990) Cysteine sulfinate (CSA) as an excitatory amino acid transmitter candidate in the mammalian central nervous system. *Prog. Neurobiol.* **35**, 313–323.

Guiramand, J., Vignes, M., and Recasens, M. (1991) A specific transduction mechanism for the glutamate action on phosphoinositide metabolism via the quisqualate metabotropic receptor in rat brain synaptoneurosomes: II. calcium dependency, cadmium inhibition. *J. Neurochem.* **57**, 1501–1509.

Haas, H. L. and Gähwiler, B. H. (1992) Vasoactive intestinal polypeptide modulates neuronal excitability in hippocampal slices of the rat. *Neurosci.* **47(2)**, 273–277.

Harvey, J. and Collingridge, G. L. (1993) Signal transduction pathways involved in the acute potentiation of NMDA responses by 1S,3R-ACPD in rat hippocampal slices. *Br. J. Pharmacol.* **109**, 1085–1090.

Heginbotham, L. R. and Dunwiddie, T. V. (1991) Long-term increases in the evoked population spike in the CA1 region of rat hippocampus induced by β-adrenergic receptor activation. *J. Neurosci.* **11(8)**, 2519–2527.

Herrero, I., Miras-Portugal, M. T., and Sanchez-Prieto, J. (1992) Positive feedback of glutamate exocytosis by metabotropic presynaptic receptor stimulation. *Nature* **360**, 163–166.

Hoehn, K. and White, T. D. (1990a) Role of excitatory amino acid receptors in K⁺-and glutamate-evoked release of endogenous adenosine from rat cortical slices. *J. Neurochem.* **54**, 256–265.

Hoehn, K. and White, T. D. (1990b) *N*-Methyl-D-aspartate, kainate, and quisqualate release endogenous adenosine from rat cortical slices. *Neurosci.* **39**, 441–450.

Holler, T., Cappel, E., Klein, J., and Löffelholz, K. (1993) Glutamate activates phospholipase D in hippocampal slices of newborn and adult rats. *J. Neurochem.* **61**, 1569–1572.

Houamed, K. M., Kuijper, J. L., Gilbert, T. L., Haldeman, B. A., O'Hara, P. J., Mulvihill, E. R., Almers, W., and Hagen, F. S. (1991) Cloning, expression, and gene structure of a G protein- coupled glutamate receptor from rat brain. *Science* **252**, 1318–1321.

Hu, G. and Storm, J. F. (1992) 2-Amino-3-phosphonopropionate fails to block postsynaptic effects of metabotropic glutamate receptors in rat hippocampal neurons. *Acta. Physiol. Scand.* **145**, 187–191.

Irving, A. J., Collingridge, G. L., and Schofield, J. G. (1992) L-Glutamate and acetylcholine mobilise Ca^{2+} from the same intracellular pool in cerebellar granule cells using transduction mechanisms with different Ca^{2+} sensitivities. *Cell Calcium* **13**, 293–301.

Irving, A. J.., Schofield, J. G., Watkins, J. C., Sunter, D. C., and Collingridge, G. L. (1990) 1S,3R-ACPD stimulates and L-AP3 blocks Ca^{2+} mobilization in rat cerebellar neurons. *Eur. J. Pharmacol.* **186**, 363–365.

Itano, Y., Murayama, T., Kitamura, Y., and Nomura, Y. (1992) Glutamate inhibits adenylate cyclase activity in dispersed rat hippocampal cells directly via an N-methyl-D-aspartate-like metabotropic receptor. *J. Neurochem.* **59**, 822–828.

Johnson, R. D. and Minneman, K. P. (1987) Differentiation of α_1-adrenergic receptors linked to phosphatidylinositol turnover and and cAMP accumulation in rat brain. *Mol. Pharmacol.* **31**, 239–246.

Knöpfel T., Vranesic I., Gähwiler B. H., and Brown D. A. (1990) Muscarinic and β-adrenergic depression of the slow Ca_{2+}-activated potassium conductance in hippocampal CA3 pyramidal cells is not mediated by a reduction of depolarization-induced cytosolic Ca^{2+} transients. *Proc. Natl. Acad. Sci. USA* **87**, 4083–4087.

Limbird, L. E. (1988) Receptors linked to inhibition of adenylate cyclase: additional signaling mechanisms. *FASEB J.* **2**, 2686–2695.

Littman, L., Glatt, B. S., and Robinson, M. B. (1993) Multiple subtypes of excitatory amino acid receptors coupled to the hydrolysis of phosphoinositides in rat brain. *J. Neurochem.* **61**, 586–593.

Löffelholz, K. (1989) Receptor regulation of choline phospholipid hydrolysis. A novel source of diacylglycerol and phosphatididic acid. *Biochem. Pharmacol.* **38**, 1543–1549.

Madison, D. V., Malenka, R. C., and Nicoll, R. A. (1991) Mechanisms underlying long-term potentiation of synaptic transmission. *Ann. Rev. Neurosci.* **14**, 379–397.

Magistretti, P. J. and Schorderet, M. (1985) Norepinephrine and histamine potentiate the increases in cyclic adenosine 3':5'-monophosphate elicited by vasoactive intestinal polypeptide in mouse cerebral cortical slices: mediation by a1-adrenergic and H1-histaminergic receptors. *J. Neurosci.* **5**, 362–368.

Manzoni, O., Prezeau, L., Rassendren, F. A., Sladeczek, F., Curry, K., and Bockaert, J. (1992a) Both enantiomers of 1-aminocyclopentyl-1,3-dicarboxylate are full agonists of metabotropic glutamate receptors coupled to phospholipase C. *Mol. Pharmacol.* **42**, 322–327.

Manzoni, O., Prezeau, L., Sladeczek, F., and Bockaert, J. (1992b) *Trans*-ACPD inhibits cAMP formation via a pertussis toxin-sensitive G-protein. *Eur. J. Pharmacol.* **225,** 357–358.

Manzoni, O., Fagni, L., Pin, J. P., Rassendren, F., Poulat, F., Sladeczek, F., and Bockaert, J. (1990) *(trans)*-1-Amino-cyclopentyl-1,3-dicarboxylate stimulates quisqualate phosphoinositide-coupled receptors but not ionotropic glutamate receptors in striatal neurons and *Xenopus* oocytes. *Mol. Pharmacol.* **38,** 1–6.

Manzoni, O. J. J., Poulat, F., Do, E., Sahuquet, A., Sassetti, I., Bockaert, J., and Sladeczek, F. A. J. (1991) Pharmacological characterization of the quisqualate receptor coupled to phospholipase C (Qp) in striatal neurons. *Eur. J. Pharmacol.* **207,** 231–241.

Masu, M. and Nakanishi, S. (1993) Molecular biology of metabotropic glutamate receptors and their physiological function. *Functional Neurology* **8,** 35.

Masu, M., Tanabe, Y., Tsuchida, K., Shigemoto, R., and Nakanishi, S. (1991) Sequence and expression of a metabotropic glutamate receptor. *Nature* **349,** 760–765.

McDonough, P. M., Goldstein, D., and Brown, J. H. (1988) Elevation of cytoplasmic calcium concentration stimulates hydrolysis of phosphatidylinositol bisphosphate in chick heart cells: effect of sodium channel activators. *Mol. Pharmacol.* **33,** 310–315.

Milani, D., Facci, L., Buso, M., Toffano, G., Leon, A., and Skaper, S. D. (1990) Excitatory amino acid receptor agonists stimulate membrane inositol phospholipid hydrolysis and increase cytoplasmic free Ca^{++} in primary cultures of retinal neurons. *Cellular Signalling* **2,** 359–368.

Murphy, S. N. and Miller, R. J. (1988) A glutamate receptor regulates Ca^{2+} mobilization in hippocampal neurons. *Proc. Natl. Acad. Sci. USA* **85,** 8737–8741.

Murphy, S. N. and Miller, R. J. (1989) Two distinct quisqualate receptors regulate Ca^{2+} homeostasis in hippocampal neurons in vitro. *Mol. Pharmacol.* **35,** 671–680.

Nakagawa, Y., Saitoh, K., Ishihara, T., Ishida, M., and Haruhiko, S. (1990) (2S,3S,4S)α-(carboxyclclopropyl)glycine is a novel agonist of metabotropic glutamate receptors. *Eur. J. Pharmacol.* **184,** 205–206.

Nakajima, Y., Iwakabe, H., Akazawa, C., Nawa, H., Shigemoto, R., Mizuno, N., and Nakanishi, S. (1993) Molecular characterization of a novel retinal metabotropic glutamate receptor mGluR6 with a high agonist selectivity for L-2-amino-4-phosphonobutyrate. *J. Biol. Chem.* **268,** 11868–11873.

Nicoletti, F., Iadarola, M. F., Wroblewski, J. T., and Costa, E. (1986a) Excitatory amino acid recognition sites coupled with inositol phospholipid metabolism: Developmental changes and interaction with α1-adrenoreceptors. *Proc. Natl. Acad. Sci. USA* **83,** 1931–1935.

Nicoletti, F., Meek, J. L., Iadarola, M. J., Chuang, D. M., Roth, B. L., and Costa, E. (1986b) Coupling of inositol phospholipid metabolism with excitatory amino acid recognition sites in rat hippocampus. *J. Neurochem.* **46,** 40–46.

Nicoletti, F., Wroblewski, J. T., Fadda, E., and Costa, E. (1988) Pertussis toxin inhibits signal transduction at a specific metabolotropic glutamate receptor in primary cultures of cerebellar granule cells. *Neuropharmacol.* **27,** 551–556.

Nicoletti, F., Magri, G., Ingrao, F., Bruno, V., Catania, M. V., Dell'Albani, P., Condorelli, D. F., and Avola, R. (1990) Excitatory amino acids stimulate inositol phospholipid hydrolysis and reduce proliferation in cultured astrocytes. *J. Neurochem.* **54,** 771–777.

Okada, D. (1992) Two pathways of cyclic GMP production through glutamate receptor-mediated nitric oxide synthesis. *J. Neurochem.* **59,** 1203–1210.

Ormandy, G. C. (1992) Inhibition of excitatory amino acid-stimulated phosphoinositide hydrolysis in rat hippocampus by L-aspartate-β-hydroxamate. *Brain Res.* **572,** 103–107.

Palmer, E., Monaghan, D. T., and Cotman, C. W. (1988) Glutamate receptors and phosphoinositide metabolism: stimulation via quisqualate receptors is inhibited by N-methyl-D-aspartate receptor activation. *Mol. Brain Res.* **4,** 161–165.

Palmer, E., Monaghan, D. T., and Cotman, C. W. (1989) Trans-ACPD, a selective agonist of the phosphoinositide-coupled excitatory amino acid receptor. *Eur. J. Pharmacol.* **166,** 585–587.

Palmer, E., Nangel-Taylor, K., Krause, J. D., Roxas, A., and Cotman, C. W. (1990) Changes in excitatory amino acid modulation of phosphoinositide metabolism during development. *Dev. Brain Res.* **51,** 132–134.

Park, T. S. and Gidday, J. M. (1990) Effect of dipyridamole on cerebral extracellular adenosine level in vivo. *J. Cereb. Blood Flow Metab.* **10,** 424–427.

Patel, J., Keith, R. A., Salama, A. I., and Moore, W. C. (1991) Role of calcium in regulation of phosphoinositide signaling pathway. *J. Mol. Neurosci.* **3,** 19–27.

Patel, J., Moore, W. C., Thompson, C., Keith, R. A., and Salama, A. I. (1990) Characterization of the quisqualate receptor linked to phosphoinositide hydrolysis in neurocortical cultures. *J. Neurochem.* **54,** 1461–1466.

Pearce, B., Morrow, C., and Murphy, S. (1990) Further characterisation of excitatory amino acid receptors coupled to phosphoinositide metabolism in astrocytes. *Neurosci. Lett.* **113,** 298–303.

Pilc, A. and Enna, S. J. (1986) Activation of a2-adrenergic receptors augments neurotransmitter-stimulated cyclic AMP accumulation in rat brain cerebral cortical slices. *J. Pharmacol. Exp. Ther.* **237,** 725–730.

Porter, P. H. P. and Roberts, P. J. (1993) Glutamate metabotropic receptor activation in neonatal rat cerebral cortex by sulphur-containing excitatory amino acids. *Neurosci. Lett.* **154,** 78–80.

Porter, R. H. P., Briggs, R. S. J., and Roberts, P. J. (1992) L-Aspartate-β-hydroxamate exhibits mixed agonist/antagonist activity at the glutamate metabotropic receptor in rat neonatal cerebrocortial slices. *Neurosci. Lett.* **144,** 87–89.

Prezeau, L., Manzoni, O., Homburger, V., Sladeczek, F., Curry, K., and Bockaert, J. (1992) Characterization of a metabotropic glutamate receptor: direct negative coupling to adenylyl cyclase and involvement of a pertussis toxin-sensitive G protein. *Proc. Natl. Acad. Sci. USA* **89,** 8040–8044.

Pullan, L. M., Olney, J. W., Price, M. T., Compton, R. P., Hood, W. F., Michel, J., and Monahan, J. B. (1987) Excitatory amino acid receptor potency and subclass specificity of sulfur-containing amino acids. *J. Neurochem.* **49,** 1301–1307.

Rana, R. S. and Hokin, L. E. (1990) Role of phosphoinositides in transmembrane signalling. *Physiological Rev.* **70,** 115–164.

Recasens, M., Guiramand, J., Nourigat, A., Sassetti, I., and Devilliers, G. (1988) A new quisqualate receptor subtype (sAA_2) responsible for the glutamate-induced inositol phosphate formation in rat brain synaptoneurosomes. *Neurochem. Int.* **13**, 463–467.

Recasens, M., Varga, V., Nanopoulos, D., Saadoun, F., Vincendon, G., and Benavides, J. (1982) Evidence for cysteine sulfinate as a neurotransmitter. *Brain Res.* **239**, 153–173.

Robertson, P. L., Bruno, G. R., and Datta, S. C. (1990) Glutamate-stimulated, guanine nucleotide-mediated phosphoinositide turnover in astrocytes is inhibited by cyclic AMP. *J. Neurochem.* **55**, 1727–1733.

Schaad, N. C., Schorderet, M., and Magistretti, P. J. (1989) Accumulation of cyclic AMP elicited by vasoactive intestinal peptide is potentiated by noradrenaline, histamine, adenosine, baclofen, phorbol esters, and ouabain in mouse cerebral slices: studies on the role of arachadonic acid metabolites and protein kinase. *J. Neurochem.* **53**, 1941–1951.

Schmidt, M. J., Thornberry, J. F., and Molloy, B. B. (1977) Effects of kainate and other glutamate analogues on cyclic nucleotide accumulation in slices of rat cerebellum. *Brain Res.* **121**, 182–189.

Schoepp, D. D. and Hillman, C. C., Jr. (1990) Developmental and pharmacological characterization of quisqualate, ibotenate and trans-1-amino-1,3-cyclopentanedicarboxylic acid stimulations of phosphoinositide hydrolysis in rat cortical brain slices. *Biogenic Amines* **7**, 331–340.

Schoepp, D. D. and Johnson, B. G. (1988) Excitatory amino acid agonist-antagonist interactions at 2-amino-4-phosphonobutyric acid-sensitive quisqualate receptors coupled to phosphoinositide hydrolysis in slices of rat hippocampus. *J. Neurochem.* **50**, 1605–1613.

Schoepp, D. D. and Johnson, B. G. (1989a) Inhibition of excitatory amino acid-stimulated phosphoinositide hydrolysis in the neonatal rat hippocampus by 2-amino-3-phosphonopropionate. *J. Neurochem.* **53**, 1865–1870.

Schoepp, D. D. and Johnson, B. G. (1989b) Comparison of excitatory amino acid-stimulated phosphoinositide hydrolysis and *N*-[³H] Acetylaspartylglutamate binding in rat brain: selective inhibition of phosphoinositide hydrolysis by 2-amino-3-phosphonopropionate. *J. Neurochem.* **53**, 273–278.

Schoepp, D. D. and Johnson, B. G. (1993) Pharmacology of metabotropic glutamate receptor inhibition of cyclic AMP formation in the adult rat hippocampus. *Neurochem. Int.* **22**, 277–283.

Schoepp, D. D., Johnson, B. G., and Monn, J. M. (1992a) Inhibition of cyclic AMP formation by a selective metabotropic glutamate receptor agonist. *J. Neurochem.* **58**, 1184–1186.

Schoepp, D. D., Johnson, B. G., Sacaan, A. I., True, R. A., and Monn, J. A. (1992b) In *vitro* and in *vivo* pharmacology of 1S,3R-and 1R,3S-ACPD: evidence for a role of metabotropic glutamate receptors in striatal motor function. *Mol. Neuropharmacol.* **2**, 33–37.

Schoepp, D. D., Johnson, B. G., True, R. A., and Monn, J. A. (1991) Comparison of (1*S*,3*R*)-1-aminocyclopentane-1,3-dicarboxylic acid (1*S*,3*R*-ACPD)-and 1*R*,3*S*-

ACPD-stimulated brain phosphoinositide hydrolysis. *Eur. J. Pharmacol. -Mol. Pharmacol.* **207,** 351–353.

Scholz, W. K. and Palfrey, H. C. (1991) Glutamate-stimulated protein phosphorylation in cultured hippocampal pyramidal neurons. *J. Neurosci.* **11(8),** 2422–2432.

Slack, J. R. and Pockett, S. (1991) Cyclic AMP induces long-term increase in synaptic efficacy in CA1 region of rat hippocampus. *Neurosci. Lett.* **130,** 69–70.

Sladeczek, F., Pin, J. P., Recasens, M., Bockaert, J., and Weiss, S. (1985) Glutamate stimulates inositol phosphate formation in striatal neurons. *Nature* **317,** 717,718.

Sortino, M. A., Nicoletti, F., and Canonico, P. L. (1991) Metabotropic glutamate receptors in rat hypothalamus characterization and developmental profile. *Dev. Brain Res.* **61,** 169–172.

Stratton, K. R., Worley, P. F., and Baraban, J. M. (1990) Pharmacological characterization of phosphoinositide-linked glutamate receptor excitation of hippocampal neurons. *Eur. J. Pharmacol.* **186,** 357–361.

Sugiyama, H., Ito, I., and Hirono, C. (1987) A new type of glutamate receptor linked to inositol phospholipid metabolism. *Nature* **325,** 531–533.

Sugiyama, H., Ito, I., and Watanabe, M. (1989) Glutamate receptor subtypes may be classified into two major categories: a study of *Xenopus* oocytes injected with rat brain mRNA. *Neuron* **3,** 129–132.

Tanabe, Y., Masu, M., Ishii, T., Shigemoto, R., and Nakanishi, S. (1992) A family of metabotropic glutamate receptors. *Neuron* **8,** 169–179.

Tanabe, Y., Nomura, A., Masu, M., Shigemotor, R., Mizuno, N., and Nakanishi, S. (1993) Signal transduction, pharmacological properties, and expression patterns of two rat metabotropic glutamate receptors, mGluR3 and mGluR4. *J. Neurosci.* **13,** 1372–1378.

Tang, W.-J. and Gilman, A. G. (1991) Type-specific regulation of adenylyl cyclase by G protein βγ subunits. *Science* **254,** 1500–1503.

Trombley, P. Q. and Westbrook, G. L. (1992) L-AP4 inhibits calcium currents and synaptic transmission via a G-protein-coupled glutamate receptor. *J. Neurosci.* **12,** 2043–2050.

Thomsen, C., Kristensen, P., Mulvihill, E., Haldeman, B., and Suzdak, P. D. (1992) L-2-amino-4-phosphonobutyrate (L-AP4) is an agonist at the type IV metabotropic glutamate receptor which is negatively coupled to adenylate cyclase. *Eur. J. Pharmacol.* **227,** 361–362.

Thomsen, C. and Suzdak, P. D. (1993) 4-carboxy-3-hydroxyphenylglycine, an antagonist at type I metabotropic glutamate receptors. *Eur. J. Pharmacol.* **245,** 299–301.

Verdoorn, T. A. and Dingledine, R. (1988) Excitatory amino acid receptors expressed in *Xenopus* oocytes: agonist pharmacology. *Mol. Pharmacol.* **34,** 298–307.

Verma, A., Hirsch, D. J., Glatt, G. V., Ronnett, G. V., and Snyder, S. H. (1993) Carbon monoxide: a putative neural messenger. *Science* **259,** 381–384.

Weiss, S. (1989) Two distinct quisqualate receptor systems are present on striatal neurons. *Brain Res.* **491,** 189–193.

Winder, D. G. and Conn, P. J. (1992) Activation of metabotropic glutamate receptors in the hippocampus increases cyclic AMP accumulation. *J. Neurochem.* **59,** 375–378.

Winder, D. G. and Conn, P. J. (1993) Activation of metabotropic glutamate receptors increases cAMP accumulation in hippocampus by potentiating responses to endogenous adenosine. *J. Neurosci.* **13,** 38–44.

Winder, D. G., Smith, T. S., and Conn, P. J. (1993) Pharmacological differentiation of metabotropic glutamate receptors coupled to potentiation of cAMP responses and phosphoinositide hydrolysis. *J. Pharmacol. Exp. Ther.* **266,** 518–525.

Wood, P. L., Emmett, M. R., Rao, T. S., Cler, J., Mick, S., and Iyengar, S. (1990) Inhibition of nitric oxide synthase blocks N-methyl-D-aspartate-, quisqualate-, kainate-, harmaline-, and pentylenetetrazole-dependent increases in cerebellar cyclic GMP in vivo. *J. Neurochem.* **55,** 346–348.

Wroblewska, B., Wroblewski, J. T., Saab, O. H., and Neale, J. H. (1993) N-acetyl-aspartylglutamate inhibits forskolin-stimulated cyclic AMP levels via a metabotropic glutamate receptor in cultured cerebellar granule cells. *J. Neurochem.* **61,** 943–948.

CHAPTER 4

Anatomical Distribution of Metabotropic Glutamate Receptors in Mammalian Brain

Claudia M. Testa, Maria Vincenza Catania, and Anne B. Young

1. Introduction

Metabotropic glutamate receptors (mGluRs) are linked to second messengers through G-proteins and play an important role in normal neuronal functions, as well as plasticity and response to injury. Until recently, it has not been possible to examine the detailed localization of mGluRs in mammalian brain because of the lack of specific agonists and antagonists amenable to ligand binding studies and the lack of cloned cDNA sequences for the development of specific *in situ* hybridization probes or antibodies for immunocytochemistry. In the past few years, many of these reagents have become available, yielding data from several avenues that have indicated marked regional heterogeneity in the distributions of mGluR subtypes. In this

The Metabotropic Glutamate Receptors Eds.: P. J. Conn and J. Patel
© 1994 Humana Press Inc., Totowa, NJ

chapter, we will review the regional expression of the five mGluR genes that have been most thoroughly characterized, the second-messenger systems associated with these receptors, and the autoradiographic binding studies of pharmacological subtypes.

2. Anatomic Distribution of mGluR Expression

2.1. Specific Subtype Expression

A family of at least seven mGluRs, named mGluR1–7, has been identified in recent molecular studies (*see* Chapter 1). This family has been divided into three groups based on the expressed clones' responses to pharmacological agents and activation of specific effector systems (Nakanishi, 1992). The first group, group 1, consists of mGluR1 and mGluR5. When expressed in cell lines, group 1 receptors appear to increase the metabolism of phosphatidylinositol *bis*-phosphate to diacylglycerol and inositol trisphosphate (IP_3), which in turn activate protein kinase C and mobilize calcium from intracellular stores, respectively (Masu et al., 1991; Abe et al., 1992; Houamed et al., 1991; *see* Chapter 3). The pharmacology of this group is unique: quisqualate is the most potent agonist for mGluR1 and mGluR5 receptors followed by glutamate. 1-Aminocyclopentane-1S,3R-dicarboxylic acid (1S,3R-ACPD) is a less potent mGluR agonist for group 1 receptors. The mGluR1 receptor has at least two splice variants (Pin et al., 1992; Tanabe et al., 1992), both in the carboxy-terminal region, one of which (mGluR1b) inserts a new stop codon, creating a much shorter coding region than the mGluR1a form. The carboxy-terminal variations may affect how the receptor couples to different second-messenger systems (Gabellini et al., 1993; Pickering et al., 1993). mGluR5 also has at least one splice variant (Minakami et al., 1993), which occurs at the same splice location as the mGluR1b variant, but does not contain a stop codon, and so most likely adds amino acids to the mGluR5a protein. The second and third groups of mGluR receptors inhibit forskolin-stimulated cAMP formation when expressed in cell lines (Tanabe et al., 1992, 1993). At the group 2 receptors, mGluR2 and mGluR3, glutamate and 1S,3R-ACPD are most potent, whereas quisqualate is a weaker agonist (Tanabe et al.,

1992, 1993). The group 3 receptor mGluR4 is unique since its most potent agonist is AP4, followed by glutamate. 1S,3R-ACPD and quisqualate are less potent at mGluR4 (Tanabe et al., 1993). Also in group 3 is mGluR6, an mGluR subtype restricted to the retina (Nakajima et al., 1993), and the more recently characterized mGluR7 (*see* Chapter 1).

Many investigators have examined the distribution of mGluR gene expression employing Northern blots (Houamed et al., 1991; Abe et al., 1992; Condorelli et al., 1992; Tanabe et al., 1992, 1993). Recent studies with *in situ* hybridization, however, give a much more detailed look at regional expression (Abe et al., 1992; Shigemoto et al., 1992; Ohishi et al., 1993a,b; Tanabe et al., 1993; Testa et al., 1994). This section will review the distribution of mGluR1–5 gene expression in rat brain. The information summarized here comes from published data using both radiolabeled riboprobes and oligonucleotides.

2.1.1. mGluR1

The expression of mGluR1 mRNA is distributed heterogeneously throughout adult brain (Shigemoto et al., 1992; Testa et al., 1994) (Fig. 1A). The region with the highest mGluR1 signal in the rat brain is the Purkinje cell layer of the cerebellum. Emulsion studies indicate heavy Purkinje cell labeling, as well as moderate labeling of granule cells. The hippocampus shows moderate levels of expression in the CA3 area and dentate gyrus, which label more strongly than the CA1 region. In addition to some CA1 pyramidal cell labeling, mGluR1 is uniquely expressed by neurons in the CA1 stratum oriens (Fotuhi et al., 1994). Also unique to mGluR1 is prominent expression in the substantia nigra pars compacta. Other structures with moderate labeling include the thalamus, olfactory bulb, septum, hypothalamus, and the subthalamic nucleus. The cerebral cortex, striatum, and cerebellar molecular layer show low levels of labeling.

2.1.2. mGluR2

The regional distribution of mGluR2 expression is very striking in rat brain, with very few regions showing consistent labeling above background on film autoradiograms (Ohishi et al., 1993a; Testa et al.,

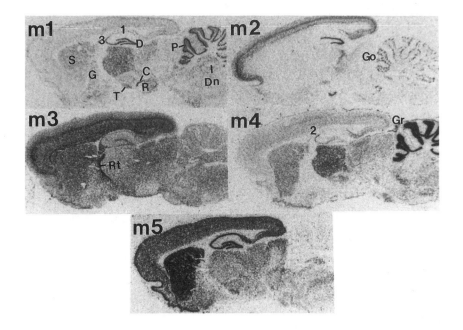

Fig. 1. Film autoradiograms of *in situ* hybridization with oligonucleotide probes to the metabotropic receptor subtype mRNAs in parasagittal sections of rat brain. The sections were prepared as described previously (Standaert et al., 1993; Testa et al., 1994). Each subtype, mGluR1 through mGluR5 (m1– m5), is depicted. 1: CA1 of hippocampus; 2: CA2 of hippocampus; 3: CA3 of hippocampus; C: pars compacta of substantia nigra; D: dentate gyrus of hippocampus; Dn: deep cerebellar nuclei; G: globus pallidus; Go: Golgi cells of cerebellum; Gr: granule cells of cerebellum; P: Purkinje cells of cerebellum; R: pars reticulata of substantia nigra; Rt: reticular nucleus of thalamus; S: striatum; T: subthalamic nucleus.

1994) (Fig. 1B). The highest level of expression is in dentate gyrus followed by Golgi cells in the granule cell layer of cerebellum. mGluR2 expression has been reported in selective thalamic, septal, amygdala, and hypothalamic nuclei. Subthalamic nucleus is also clearly labeled, and the cerebral cortex has a moderate level of expression with distinct lamination. On examination of emulsion-dipped slides, cortical labeling is most dense in layer IV granule cells, although pyramidal cells in layers III and V are also labeled.

Although very little labeling in the basal ganglia is evident on film autoradiograms, in emulsion studies, a small population of striatal neurons consisting of large, presumably cholinergic neurons is clearly labeled.

2.1.3. mGluR3

mGluR3 labeling is prominent and diffusely distributed through-out mammalian brain (Ohishi et al., 1993b; Tanabe et al., 1993; Testa et al., 1994) (Fig. 1C). Of interest is the strong labeling of glial cells as well as neurons. Labeling of white matter is obvious on film auto-radiograms, and on examination of emulsion-dipped slides, labeling is clearly present over both neuronal and glial profiles. The most heavily labeled structure within the mammalian brain is the reticular nucleus of the thalamus. Neurons and glia within this structure are densely labeled. The cerebral cortex also exhibits dense labeling, particularly over layer V pyramidal cells. Layer III pyramidal cells are also labeled. In the cerebellum, labeling is modest, but at the emulsion level, strong labeling over stellate and basket cells can be seen. Striatal labeling is quite prominent and, as in reticular nucleus, both neurons and glia express this gene. Neuronal labeling has also been reported in the inferior and superior colliculi, as well as some amygdala and hypothalamic nuclei.

2.1.4. mGluR4

mGluR4, like mGluR2, has a very restricted distribution in adult brain (Tanabe et al., 1993; Testa et al., 1994) (Fig. 1D). mGluR4 expression is four times higher in the granule cell layer of cerebellum than in any other region. The next highest level of labeling is in the ventral nuclei of the thalamus and the olfactory bulb. There is rela-tively low labeling of hippocampal structures, but a striking finding is a clear definition of the CA2 region pyramidal cells with this recep-tor probe (Fotuhi et al., 1994). The striatum and nucleus accumbens are also modestly labeled.

2.1.5. mGluR5

As with mGluR3 and mGluR1, mGluR5 expression is observed throughout mammalian brain (Abe et al., 1992; Testa et al., 1994) (Fig. 1E). The striatum is strikingly labeled in a heterogeneous pattern. Emulsion examination reveals that about three-quarters of the dorsolateral striatal neurons are strongly labeled, whereas the remaining 25% have little or no consistent labeling (Testa et al., 1994). The large polygonal neurons, which are labeled with probes to mGluR2, appear distinctly unlabeled with the mGluR5 probes. Hippocampus is also heavily labeled for mGluR5. Labeling is equally intense in dentate gyrus, CA3, and CA1. In clear contrast to mGluR4, mGluR5 expression is high in forebrain, but very low in cerebellum. The cerebellar cortex is nearly devoid of labeling for mGluR5, but there is discernible signal in the deep cerebellar nuclei. Cerebral cortex and subthalamic nucleus also show modest labeling.

2.2. Regional Subtype Expression

Metabotropic receptors have been implicated in the normal functioning and pathology of several central nervous system regions. Knowledge of the distinct subtype expression patterns in these regions is crucial to the understanding of the specific mechanisms involved in mGluR-mediated processes.

2.2.1. Hippocampus

The expression of mGluRs in hippocampus is of interest because of the role these receptors play in long-term potentiation (LTP), a model of learning (Aronica et al., 1991; Bashir et al., 1993; Bliss and Collingridge, 1993), and in neuronal injury (Choi and Rothman, 1990; Sacaan and Schoepp, 1992; Seren et al., 1989). Also, mGluR stimulation in rat hippocampal slices induces a number of second-messenger and electrophysiological effects (*see* Chapter 8). The mGluR subtypes responsible for these effects will need to be defined with detailed information about subtype expression as well as physiology. mGluR1 is expressed most prominently in the dentate gyrus granule cells and CA3, whereas mGluR2 is enriched in the dentate gyrus and the inner layer of entorhinal cortex (Figs. 2 and 3). mGluR3 is also present in these two structures, but unlike the other mGluRs, it is found in glia as well as neurons. As stated above, mGluR4 is present predominantly in CA2, whereas mGluR5 is expressed in high density throughout most regions of hippocampus and entorhinal cortex.

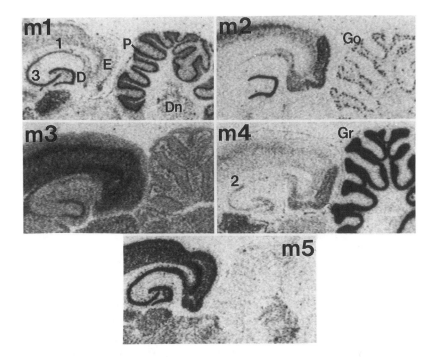

Fig. 2. Film autoradiograms of *in situ* hybridization of metabotropic receptor mRNAs in the hippocampus and cerebellum in horizontal sections processed in parallel with the parasagittal sections shown in Fig. 1. Each subtype of metabotropic glutamate receptor, mGluR1 through mGluR5 (m1–m5), is depicted. Structures are the same as labeled in Fig. 1 with the addition of E: entorhinal cortex.

The distribution of mGluR5 suggests that mGluR5 may be the primary subtype of metabotropic receptor mediating the excitatory actions of glutamate on CA1 (Fotuhi et al., 1994), and could therefore contribute to the elevation of calcium levels found in CA1 pyramidal neurons during long-term potentiation (Bliss and Collingridge, 1993) or after hypoxic ischemic injury (Choi and Rothman, 1990). mGluR2 and mGluR3 are absent in CA1. In fact, mGluR2 does not appear to be present in hippocampal pyramidal cells; instead, it may be located on presynaptic terminals of dentate gyrus granule cells synapsing on CA3 pyramidal neurons and on presynaptic terminals of entorhinal

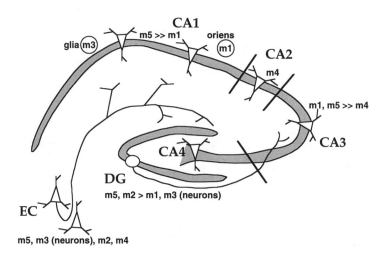

Fig. 3. Schematic diagram of the cellular localization of metabotropic glutamate receptor gene expression in the hippocampal formation. Messenger RNA for each of the mGluR1–5 receptors (designated m1–5 in the figure) is indicated in the region of the cell bodies in which they are found. Protein localizations may be postulated from this data; for example, mGluR2 receptors may be present in entorhinal cortex and dentate gyrus granule cell projection terminals. Details of pre- and postsynaptic localizations await further studies. The amounts of the receptor subtypes indicated are based on comparison of relative *in situ* hybridization signal levels, not quantitative measurement of mRNA levels. DG: dentate gyrus; EC: entorhinal cortex; oriens: stratum oriens of CA1; CA1, CA2, CA3, and CA4 delineate the standard subdivisions of the Ammon's horn.

cortex neurons projecting to hippocampus. These distributions suggest that mGluR-mediated excitatory actions of glutamate can occur in all regions of hippocampus, whereas inhibitory actions may be restricted to dentate gyrus or presynaptic terminals (Fotuhi et al., 1994).

2.2.2. Cerebellum

A model of synaptic plasticity similar to hippocampal LTP exists in the cerebellum (*see* Chapters 7 and 8). Long-term depression (LTD), a process underlying some forms of motor learning, can be induced at the parallel fiber-Purkinje cell synapse by concurrent activation of climbing fiber and parallel fiber inputs to a Purkinje cell (Ito et al., 1982). It has been shown that a postsynaptic mGluR on Purkinje cell dendrites is required for LTD (Kano and Kato, 1987; Linden et al., 1991). Growing evidence, including physiological (Linden et al., 1991), *in situ* hybridization (Shigemoto et al., 1992), and immunohistochemistry (Martin et al., 1992; Fotuhi et al., 1993), indicates that mGluR1 mediates LTD in the cerebellum.

Although cells in the cerebellum may express several different mGluRs, nearly all cerebellar cell types have a single predominant mGluR (Figs. 2 and 4). In addition, the hybridization signal of the major mGluR in a cerebellar cell type is often the highest signal for that receptor message in adult brain (Abe et al., 1992; Shigemoto et al., 1992; Ohishi et al., 1993a,b; Tanabe et al., 1993; Testa et al., 1994). Labeling for mGluR1 is highest in Purkinje cells, where mGluR1 receptors may contribute an important component to synaptic plasticity as discussed above. mGluR2 expression in cerebellum is observed exclusively in Golgi cells, at a level comparable only to that of mGluR2 in the dentate gyrus of the hippocampus. Golgi cells also appear to express low levels of mGluR5, mGluR1, and mGluR3, but the clear demarcation of this cell type by mGluR2 expression suggests a unique physiological role for mGluR2 receptors in Golgi cell function. Granule cells express the highest levels of mGluR4 in the adult brain. Again, although low levels of other receptor signals, such as mGluR1, are observed in granule cells, mGluR4 is likely to play a major functional role in the biology of this cell type. mGluR3 is much more prominent in stellate and basket cells than in other cerebellar cell types. Finally, mGluR5 is seen at very low levels in Golgi cells, but is most notable for its absence from the cerebellar cortex. Instead, mGluR5 is most prominent in the deep cerebellar nuclei.

2.2.3. Basal Ganglia

Metabotropic receptors are thought to play an important role in basal ganglia function. The basal ganglia are a group of deep forebrain nuclei that regulate motor behavior and play a central function in human diseases, such as Parkinson's and Huntington's disease. The major excitatory input to the basal ganglia comes from the cerebral cortex and is likely to use glutamate as a neurotransmitter. This

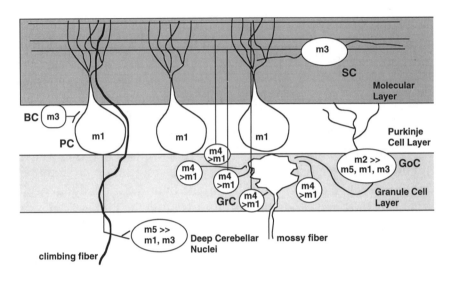

Fig. 4. Schematic diagram of the cellular localization of metabotropic glutamate receptor gene expression in cerebellum. The relative gene expression in each cell is indicated for mGluR1–5 (designated m1–5 in the figure). The data represented are from comparison of relative *in situ* hybridization signal levels, not quantitative measurement of mRNA levels. BC: basket cell; GoC: Golgi cell; GrC: granule cell; PC: Purkinje cell; SC: stellate cell.

pathway has been extensively studied biochemically, physiologically, and behaviorally (Kim et al., 1977; McGeer et al., 1977; Young et al., 1981). The subthalamic nucleus and the substantia nigra also receive excitatory input from cerebral cortex (Beckstead, 1979; Lee et al., 1988; Canteras et al., 1990; Groenewegen and Berendse, 1990). The subthalamic nucleus itself is thought to be glutamatergic and projects to substantia nigra, pallidum and neostriatum (Kita and Kitai, 1987; Nakanishi et al., 1987; Smith and Parent, 1988; Albin et al., 1989; Groenewegen and Berendse, 1990; Robledo and Feger, 1990; Brotchie and Crossman, 1991; Price et al., 1993). The subthalamic nucleus is thought to play a key role in regulating the final output of the basal ganglia (Albin et al., 1989).

In the striatum, all of the mGluRs are expressed with widely differing intensities and cellular patterns (Testa et al., 1994) (Figs. 1

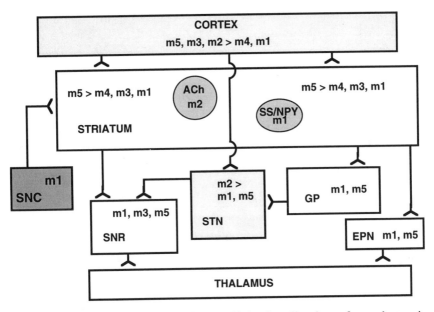

Fig. 5. Schematic diagram of the cellular localization of metabotropic glutamate receptor gene expression in the basal ganglia. The relative subtype expression in each area is indicated for mGluR1–5 (designated m1–5 in the figure). The amounts of the receptor subtypes indicated are based on comparison of relative *in situ* hybridization signal levels, not quantitative measurement of mRNA levels. All the glutamatergic regions are lightly stippled; GABAergic nuclei are clear; the dopaminergic nucleus is densely stippled. Striatal interneurons are moderately stippled. ACh: acetylcholine interneurons; EPN: entopeduncular nucleus; GP: globus pallidus; SNC: substantia nigra pars compacta; SNR: substantia nigra pars reticulata; SS/NPY: somatostatin/neuropeptide Y interneurons; STN: subthalamic nucleus.

and 5). The most prominent labeling is for mGluR5, which shows a heterogeneous cellular distribution, labeling a large subpopulation of medium-sized striatal neurons. mGluR1 is found at low levels throughout the striatum. Some studies suggest that mGluRs in the striatum may enhance glutamate neurotoxicity (Aleppo et al., 1992; McDonald and Schoepp, 1992; Sacaan and Schoepp, 1992; Beal et al., 1993). Since excitotoxicity is thought to require an increase in intracellular calcium, the phospholipase C-coupled group 1 receptors mGluR1 and mGluR5, particularly the abundant mGluR5, are potential candidates for the mGluR subtypes with such an enhancing role in striatal excitotoxicity.

Striatal metabotropic receptors are involved in a form of LTD similar to that seen in the cerebellum. Striatal LTD can be produced by tetanic stimulation of corticostriatal fibers in a slice preparation, requires dopamine receptor activation, and is blocked by an mGluR antagonist (Calabresi et al., 1992). mGluR1, which is expressed in nearly all striatal neurons, could mediate striatal LTD through a postsynaptic receptor similar to the one implicated in cerebellar LTD.

mGluR3 is expressed in neurons and glia in the striatum, whereas mGluR4 is found in a majority of medium-sized striatal neurons and mGluR2 is seen only in a small population of large polygonal, presumably cholinergic interneurons (Testa et al., 1994). The potential for different striatal cells types (cholinergic interneurons, somatostatin interneurons, populations of projection neurons) to have distinct profiles of mGluR expression creates opportunities to target specific groups of neurons or second-messenger effects in the study and therapy of motor regulation. For example, mGluR5 and mGluR4 exhibit selective expression in subpopulations of medium-sized striatal neurons, whereas large polygonal neurons appear to only express mGluR2, suggesting that agonists specific for these receptors may have important, differential effects on motor regulation.

Elsewhere in the basal ganglia, mGluR3 is expressed in glia in all basal ganglia structures (Ohishi et al., 1993b), as well as in a subset of neurons in substantia nigra pars reticulata (Testa et al., 1994). mGluR1 is the only mGluR prominently expressed by the dopaminergic neurons of the substantia nigra pars compacta (Shigemoto et al., 1992; Testa et al., 1994). mGluR2 is expressed only in neurons of the subthalamic nucleus (Ohishi et al., 1993a; Testa et al., 1994), whereas there is little significant mGluR4 expression outside the striatum and thalamus (Tanabe et al., 1993; Testa et al., 1994). In addition to mGluR2, subthalamic nucleus neurons also express mGluR5 and mGluR1 (Testa et al., 1994). Overactivity of the subthalamic nucleus has a potentially important role in the clinical syndrome of Parkinson's disease (Albin et al., 1989). The profile of subthalamic nucleus mGluRs, particularly the unique expression of mGluR2, suggests that mGluRs mediate inhibitory as well as excitatory glutamatergic actions in this nucleus, and that agents targeted to mGluR2 may provide a pharmacological means of selectively altering subthalamic nucleus function.

A variety of mechanisms may underly complex mGluR-mediated motor behaviors. The differential distribution of mGluRs throughout the basal ganglia can indicate a set of distinct potential roles in a given system. For example, unilateral injection of the mGluR agonist 1S,3R-ACPD into the striatum of rats induces contralateral rotational behavior via a chain of events that includes increased release of dopamine (Sacaan et al., 1991, 1992). The location and type of mGluRs that mediate this effect could include postsynaptic mGluRs on different striatal neuronal populations as discussed above. Alternatively, presynaptic mGluR1 receptors on nigrostriatal dopaminergic terminals could play a role, or presynaptic mGluR2 or mGluR3 receptors on corticostriatal terminals may be the principal mediators of this effect. In studies to elucidate the individual roles of mGluR subtypes in this system, the distinct mGluR subtype expression patterns in basal ganglia will allow targeting of specific mGluRs anatomically and pharmacologically.

3. Anatomy of Metabotropic Receptor-Linked Second-Messenger Systems

3.1. Phosphoinositol Metabolism

The second-messenger system that was first associated with mGluRs is the metabolism of phosphatidylinositol (PI) (Sladeczak et al., 1985; *see* Chapter 3). The turnover of PI has been used as a measure of mGluR activity in a number of systems (Sladeczak et al., 1985; Nicoletti et al., 1986, 1987; Sugiyama et al., 1987). With the recent cloning of the family of mGluRs, it has become clear that only two of several known mGluRs are coupled to PI turnover (Abe et al., 1992; Tanabe et al., 1992). The group 1 receptors mGluR1 and mGluR5 appear to use different G-proteins to affect PI metabolism, since mGluR1-mediated PI turnover is sensitive to pertussis toxin blockade, whereas mGluR5-mediated PI turnover is not (Abe et al.,

1992). Using [³H]cytidine as a precursor, it is possible to visualize the anatomical distribution of PI turnover by using autoradiography to detect a membrane-bound product containing the labeled cytidine (Hwang et al., 1990).

PI is metabolized to diacylglycerol, which activates protein kinase C (PKC), and IP_3, which in turn acts through the IP_3 receptor to release calcium from intracellular stores. PKC and IP_3 receptors are differentially distributed in rat brain (Worley et al., 1986a,b, 1989); thus, activation of mGluR1 or mGluR5 may have very different consequences in structures that have different components of the PI second-messenger system available.

3.1.1. IP_3 Receptor Binding

IP_3 receptors can be mapped using [³H]IP_3 receptor autoradiography (Worley et al., 1989), immunohistochemistry (Fotuhi et al., 1993) and *in situ* hybridization (Fotuhi et al., 1993). IP_3 receptors are most abundant in the molecular layer of the cerebellum, where they are located in Purkinje cell dendrites, the proposed location of mGluR1. In the hippocampus, IP_3 receptors are selectively localized in CA1 and dentate gyrus, whereas mGluR5 expression is high throughout the hippocampus; therefore, mGluR5 may affect calcium levels via IP_3 receptors in CA1 and dentate, but act through a different system in CA3. IP_3 receptors are also present in cerebral cortex, striatum, and the substantia nigra, possibly in striatonigral terminals.

3.1.2. Phorbol Ester Binding

The second branch of the PI second-messenger system, PKC activation by diacylglycerol, can be localized using autoradiography of [³H]phorbol 12,13-dibutyrate ([³H]PDBu) binding (Worley et al., 1986a,b). [³H]PDBu binding is most abundant in rat brain in intrinsic neurons of cerebral cortex, and stratum oriens and stratum radiatum layers of CA1 of the hippocampus where mGluR1 expression is observed (Fotuhi et al., 1994; Testa et al., 1994); binding is also high in these layers in CA3 and in the molecular layer of the dentate gyrus. There is a high level of [³H]PDBu binding in the cerebellar molecular layer, where it is mostly present in dendrites of Purkinje cells, again colocalizing with mGluR1. In the basal ganglia, the striatum and substantia nigra contain high levels of [^{3}H]PDBu binding; thus, both branches of the PI pathway are available for mGluR1 and mGluR5 actions in these nuclei. Binding in the substantia nigra may represent PKC levels in striatonigral projection terminals.

3.1.3. Phophoinositide Turnover Autoradiography

[^{3}H]Cytidine is incorporated during PI turnover into [^{3}H]cytidine diphosphate diacylglycerol ([^{3}H]CDP-DAG), which remains membrane bound (Hwang et al., 1990). 1S,3R-ACPD can be used to stimulate PI turnover in the presence of [^{3}H]cytidine. Subsequent autoradiography reveals [^{3}H]CDP-DAG accumulation in cerebellar molecular and Purkinje cell layers, CA3 of hippocampus, and in subiculum (Hwang et al., 1990; Fotuhi et al., 1993).

3.2. Adenylyl Cyclase-Linked Functions

Metabotropic receptors are capable of either increasing or decreasing cAMP formation by adenylyl cyclase (*see* Chapter 3). mGluR1 expressed in CHO cells increases cAMP formation (Aramori and Nakanishi, 1992), though mGluR5 does not (Abe et al., 1992). An mGluR of unknown type in the hippocampus can also increase cAMP formation in hippocampal slices, via a synergistic action with other G$_s$-coupled receptors (Winder and Conn, 1993). In contrast, the group 2 and group 3 receptors, mGluR2–4, decrease forskolin-stimulated cAMP formation in vitro (Hayashi et al., 1992; Tanabe et al., 1992, 1993), creating a potential new inhibitory mechanism of glutamate neurotransmission.

3.2.1. Forskolin Binding

The adenylyl cyclase second-messenger system can be mapped using [^{3}H]forskolin, which binds displaceably to the adenylyl cyclase–G$_s$ complex (Worley et al., 1986a). In the hippocampus, [^{3}H]forskolin binding is abundant on intrinsic neurons of the dentate gyrus molecular layer and the pyramidal layer of CA3 and CA4, all areas that express mGluR1–3. The highest levels of [^{3}H]forskolin binding occur in the cerebellar molecular layer, striatum, nucleus accumbens, and substantia nigra. In the cerebellum, binding is associated with granule cells, which also have the highest signal levels for mGluR4 mRNA. [³H]Forskolin binding in the substantia nigra appears to be associated with striatonigral terminals, like [³H]PDBu binding.

4. Distribution of mGluR Proteins

4.1. Autoradiography of mGluR Binding

Quantitative receptor autoradiographic techniques allow the measurement of presumed receptor proteins by measuring the amount of radiolabeled ligand bound to tissue sections under conditions favoring binding to a specific receptor (Young and Fagg, 1990). Using an unlabeled competitor to displace a radiolabeled ligand gives an apparent affinity and number of binding sites for the competitor. Multiple sets of competitor binding sites with different affinities can also be revealed by this method.

The identification of ligand binding sites depends in part on the choice of radiolabeled ligand, the use of saturating levels of competitors for other sites to restrict the range of sites available to the radiolabeled ligand, and the pharmacological profile of the resultant binding. [³H]Glutamate is a commonly used ligand for glutamate receptor binding studies. The addition of high concentrations of AMPA, NMDA, and calcium chloride to the binding mixture saturates the ionotropic glutamate receptors, leaving only a subset of glutamate receptors available for examination (Cha et al., 1990; Greenamyre et al., 1990). [³H]Glutamate binding under these conditions displays a pharmacological profile indicative of mGluRs: it is displaceable by quisqualate and 1S,3R-ACPD (Cha et al., 1990; Catania et al., 1993) but not 1R,3S-ACPD (Catania et al., 1993), and the binding affinity is decreased by GTPγS, a guanine nucleotide binding analog that disrupts coupling of receptors to G-proteins (Catania et al., 1994). In addition, quisqualate displacement of metabotropic-like [³H]glutamate binding yields a biphasic curve, indicating the existence of two different populations of binding sites with high (nanomolar) and low (micromolar) affinities for quisqualate (Catania et al., 1993). The quisqualate displacement data have been used to define two different metabotropic-like binding sites: met1, a high-affinity quisqualate site, and met2, the low-affinity quisqualate site. 1S,3R-ACPD selectively displaces binding under both met1 and met2 conditions, suggesting that both binding site types represent binding to mGluR proteins (Catania et al., 1993, 1994).

4.1.1. High-Affinity Quisqualate Sites

The high-affinity quisqualate binding is studied as the difference in [³H]glutamate binding under mGluR-favoring conditions in the absence and presence of 2.5 μ*M* quisqualate (Cha et al., 1990; Catania et al., 1993). The site with this profile, met1, is found throughout the rat central nervous system (Cha et al., 1990; Albin et al., 1992; Catania et al., 1993, 1994). The highest levels of met1 are in the cerebellar molecular layer, septum, stratum moleculare of the dentate gyrus, and stratum oriens of the CA1 of the hippocampus, as well as the entorhinal cortex (Fig. 6). Met1 sites are also present in the striatum, globus pallidus, ventral pallidus, stratum radiatum of CA1 and CA3 of the hippocampus, and throughout the thalamus, though met1 binding is virtually absent in the reticular nucleus.

The pharmacological profile of met1 binding is similar to that of the group 1 mGluR clones, mGluR1 and mGluR5 (Masu et al., 1991; Abe et al., 1992; Nakanishi, 1992; Catania et al., 1993, 1994). The regional distribution of met1 sites closely corresponds to the distributions of mGluR1 and mGluR5 mRNA expression (Testa et al., 1994). In the basal ganglia, met1 binding is high in the striatum where mGluR5 is strongly expressed, though mGluR1 is present only at modest levels. The relative levels of met1 binding across the basal ganglia nuclei—pallidus, subthalamic nucleus, substantia nigra—are the same as those for mGluR1 and mGluR5 expression, with mGluR1 relatively more prominent in these nuclei. The high levels of met1 sites in the cerebellar molecular layer may reflect binding to postsynaptic mGluR1 receptors on Purkinje cell dendrites. Met1 binding is seen throughout the hippocampus. mGluR1 expression in the stratum oriens of CA1 may be related to met1 binding in this area. mGluR1 expression is also prominent in the pyramidal

Fig. 6. Autoradiographic images of [³H]glutamate binding to met1 and met2 binding sites in rodent brain. Left: Total [³H]glutamate binding in the presence of 50 m*M* Tris-HCl, 2.5 m*M* CaCl2, 30 m*M* potassium thiocyanate, 100 μ*M* NMDA, 100 μ*M* AMPA. Right: [³H]Glutamate binding in the same conditions as for left image except with the addition of 2.5 μ*M* quisqualate. The binding shown is displaceable by 1S,3R-ACPD and has a distribution consistent with binding to the metabotropic receptors that inhibit forskolin-stimulated adenylyl cyclase (the met2 site). This binding site is not sensitive to AP4, which acts at mGluR4 receptors. Middle: Image of binding that remains after subtracting the image on the right from the image on the left. This image is likely to represent binding to met1 receptors linked to phosphatidyl inositol turnover.

layer of CA3 and the granular layer of dentate gyrus, whereas mGluR5 is found in all hippocampal areas.

4.1.2. Low-Affinity Quisqualate Sites

Met2 binding is measured as the difference between [³H]gluta-mate binding under mGluR-favoring conditions with 2.5 μ*M* quisqualate to saturate met1 sites in the absence and presence of 1S,3R-ACPD. This low-affinity quisqualate binding has a more prominent differential distribution in rat brain than the high-affinity quisqualate site (Greenamyre et al., 1990; Albin et al., 1992; Catania et al., 1993, 1994). Met2 binding is most prominent in the outer cerebral cortex (layers I–III) (Fig. 6). Met2 sites are also present in the striatum, septum, thalamus including the reticular nucleus, and stratum moleculare of all areas of the hippocampus, but are very low in the cerebellum, where met1 sites are abundant.

The met2 site pharmacology resembles that of the group 2 receptors, mGluR2 and mGluR3 (Tanabe et al., 1992, 1993; Catania et al., 1994). In some areas, the regional distributions of these transcripts match that of the met2 binding (Testa et al., 1994); for example, both met2 and mGluR3 expression are uniquely abundant in the reticular nucleus of the thalamus. In addition, mGluR2 is highly expressed in the granular layer of dentate gyrus of the hippocampus. The dendrites of these cells are located in the stratum moleculare of the dentate gyrus, an area with high met2 binding, but low met1 binding. In other areas, met2 ligand binding and group 2 mGluR expression appear unrelated.

Some of the apparent discordance of ligand binding sites and mGluR gene expression may be explained by the potential presynaptic localization of mGluRs. For example, high mGluR3 signal and mGluR2 expression are observed in cortical pyramidal neurons, which project to the striatum, an area of high met2 binding but relatively low group 2 transcript expression. Presynaptic mGluR2 (Ohishi et al., 1993a; Testa et al., 1994) and/or mGluR3 receptors could be present on corticostriatal terminals. mGluR2 is also present at moderate levels in the subthalamic nucleus, where met2 binding is low, but is virtually absent in subthalamic projection targets where met2 binding is moderate (pallidal and nigral nuclei) or high (striatum), suggesting the existence of presynaptic mGluR2 receptors on subthalamic nucleus projection terminals (Testa et al., 1994).

In other areas, evidence exists for both pre- and postsynaptic group 2 receptors. Met2 binding, unlike met1, is high in the stratum moleculare of all areas of the hippocampus. Of the mGluRs, only mGluR2 is expressed at high levels in the inner entorhinal cortex, which sends projections to CA1 and particularly to CA3 stratum moleculare, arguing for presynaptic mGluR2 receptors on these axons. On the other hand, mGluR2 is also expressed intensely by the granular layer cells of the dentate gyrus, which have dendrites in the stratum moleculare of dentate gyrus and send axonal projections to stratum lucidum of CA3. The high met2 binding in dentate stratum moleculare suggests the presence of postsynaptic mGluR2 receptors on granular layer dendrites, though some met2 binding is also present in CA3 stratum lucidum.

4.2. Immunocytochemistry of mGluRs

Currently the only published antibodies available for immuno-cytochemical localization of specific mGluR-subtype proteins are directed against mGluR1. These include an antibody directed to an amino-terminal area shared by all three known mGluR1 splice variants (Fotuhi et al., 1993) and another antibody specific for the carboxy-terminal of the mGluR1a splice variant (Martin et al., 1992). mGluR1 protein is found at high levels in the cerebellum, hippocampus, thalamus, and olfactory bulb, and is also present in many other structures, including cerebral cortex, striatum, and substantia nigra. In the cerebellum, mGluR1a is present in Purkinje cells bodies and dendritic shafts (Martin et al., 1992), but is not seen in granule cells, which are immunopositive with the general mGluR1 antibody (Fotuhi et al., 1993). In the hippocampus, mGluR1 is high in stratum oriens of CA1 and the molecular layer of dentate gyrus.

Differences between the distribution of mGluR1 mRNA, receptor binding, and immunoreactivity with mGluR1- or mGluR1a-selective antibodies suggest a presynaptic location for certain forms of mGluR1. The distribution of mGluR1 in the basal ganglia closely follows the mRNA expression pattern, except for low protein levels in the substantia nigra pars compacta (Martin et al., 1992; Fotuhi et al., 1993), where mRNA signal is relatively high (Shigemoto et al., 1992; Testa et al., 1994). The general mGluR1 antibody, but not the mGluR1a antibody, detects high protein levels in presynaptic terminals in the striatum, possibly reflecting mGluR1b receptors on nigrostriatal terminals. In addition, staining of presynaptic terminal fields is seen within the cerebral cortex with the general mGluR1 antibody only, possibly because of presynaptic mGluR1 receptors on thalamocortical fibers. Since the carboxy-terminal splicing of mGluR1 may regulate receptor coupling to second-messenger systems (Gabellini et al., 1993; Pickering et al., 1993), these anatomical results indicate possible different physiological responses of mGluR1 receptors at pre- vs postsynaptic sites.

5. Summary

Considerable information is now available on the regional localization of metabotropic glutamate receptors; however, new receptors and splice variants are being identified at a rapid rate. In the near future, much more detailed data at the light and electron microscopic levels using gene probes and selective antibodies will shed additional light on the selective actions of these receptors.

Acknowledgments

This work was supported by USPHS grants NS 19613 and AG 08671, and a Medical Scientist Training Program Fellow award to C. M. T.

References

Abe, T., Sugihara, H., Nawa, H., Shigemoto, R., Mizuno, N., and Nakanishi, S. (1992) Molecular characterization of a novel metabotropic glutamate receptor, mGluR5 coupled to inositol phosphate/Ca^{2+} signal transduction. *J. Biol. Chem.* **267,** 13,361–13,368.

Albin, R. L., Young, A. B., and Penney, J. B. (1989) The functional anatomy of basal ganglia disorders. *Trends Neurosci.* **12,** 366–375.

Albin, R. L., Aldridge, J. W., Young, A. B., and Gilman, S. (1989) Feline subthalamic nucleus neurons contain glutamate-like but not GABA-like or glycine-like immunoreactivity. *Brain Res.* **491,** 185–188.

Albin, R. L., Makowiec, R. L., Hollingsworth, Z., Dure, L. S., Penney, J. B., and Young, A. B. (1992) Excitatory amino acid binding sites in the basal ganglia of the rat: a quantitative autoradiographic study. *Neurosci.* **46,** 35–48.

Aleppo, G., Pisani, A., Copani, A., Bruno, V., Aronica, E., D'Agata, V., Canonico, P. L., and Nicoletti F. (1992) Metabotropic glutamate receptors and neuronal toxicity. *Adv. Exp. Med. Biol.* **318,** 137–145.

Aramori, I. and Nakanishi, S. (1992) Signal transduction and pharmacological characteristics of a metabotropic glutamate receptor, mGluR1, in transfected CHO cells. *Neuron* **8,** 757–765.

Aronica, E., Frey, U., Wagner, M., Schroeder, H., Krug, M., Ruthrich, H., Catania, M. V., Nicoletti, F., and Reymann, K. G. (1991) Enhanced sensitivity of "metabotropic" glutamate receptors after induction of long-term potentiation in rat hippocampus. *J. Neurochem.* **57,** 376–383.

Bashir, Z. I., Bortolotto, Z. A., Davies, C. H., Berretta, N., Irving, A. J., Seal, A. J., Henley, J. M., Jane, D. E., Watkins, J. C., and Collingridge, G. L. (1993) Induction of LTP in the hippocampus needs synaptic activation of glutamate metabotropic receptors. *Nature* **363**, 347–350.

Beal, M. F., Finn, S. F., and Brouillet, E. (1993) Evidence for the involvement of metabotropic glutamate receptors in striatal excitotoxin lesions *in vivo*. *Neurodegen.* **2**, 81–91.

Beckstead, R. M. (1979) An autoradiographic examination of corticocortical and subcortical projections of the mediodorsal-projection (prefrontal) cortex in the rat. *J. Comp. Neurol.* **184**, 43–62.

Bliss, T. V. P. and Collingridge, G. L. (1993) A synaptic model of memory: long-term potentiation in the hippocampus. *Nature* **361**, 31–39.

Brotchie, J. M. and Crossman, A. R. (1991) D-[³H]Aspartate and [¹⁴C]GABA uptake in the basal ganglia of rats following lesions in the subthalamic region suggest a role for excitatory amino acid but not GABA-mediated transmission in subthalamic nucleus efferents. *Exp. Neurol.* **113**, 171–181.

Calabresi, P., Maj, R., Pisani, A., Mercuri, N. B., and Bernardi G. (1992) Long-term synaptic depression in the striatum: physiological and pharmacological characterization. *J. Neurosci.* **12**, 4224–4233.

Canteras, N. S., Shammah-Lagnado, S. J., Silva, B. A., and Ricardo, J. A. (1990) Afferent connections of the subthalamic nucleus: A combined retrograde and anterograde horseradish peroxidase study in the rat. *Brain Res.* **513**, 43–59.

Catania, M. V., de Socarraz, H., Penney, J. B., and Young, A. B. (1994) Metabotropic glutamate receptor heterogeneity in rat brain. *Mol. Pharmcol.* (in press).

Catania, M. V., Hollingsworth, Z., Penney, J. B., and Young, A. B. (1993) Quisqualate resolves two distinct metabotropic [³H]glutamate binding sites. *NeuroRep.* **4**, 311–313.

Cha, J. H., Makowiec, R. L., Penney, J. B., and Young, A. B. (1990) L-[³H]glutamate labels the metabotropic excitatory amino acid receptor in rodent brain. *Neurosci. Let.* **113**, 78–83.

Choi, D. W. and Rothman, S. M. (1990) The role of glutamate neurotoxicity in hypoxic-ischemic neuronal death. *Ann. Rev. Neurosci.* **13**, 171–182.

Condorelli, D. F., Dell'Albani, P., Amico, C., Casabona, G., Genazzani, A. A., Sortino, M. A., and Nicoletti, F. (1992) Developmental profile of metabotropic glutamate receptor mRNA in rat brain. *Mol. Pharmacol.* **41**, 660–664.

Fotuhi, M., Sharp, A. H., Glatt, C. E., Hwang, P. M., von Krosigk, M., Snyder, S. H., and Dawson, T. M. (1993) Differential localization of phosphoinositide-linked metabotropic glutamate receptor (mGluR1) and the inositol 1,4,5-trisphosphate receptor in rat brain. *J. Neurosci.* **13**, 2001–2012.

Fotuhi, M., Standaert, D. G., Testa, C. M., Penney, J. B., and Young, A. B. (1994) Differential expression of metabotropic glutamate receptors in the hippocampus and entorhinal cortex of the rat. *Mol. Brain Res.* **21**, 283–292.

Gabellini, N., Manev, R. M., Candeo, P., Favaron, M., and Manev, H. (1993) Carboxyl domain of glutamate receptor directs its coupling to metabolic pathways. *NeuroRep.* **4**, 531–534.

Greenamyre, J. T., Higgins, D. S., Young, A. B., and Penney, J. B. (1990) Regional ontogeny of a unique glutamate recognition site in rat brain: an autoradiographic study. *Int. J. Dev. Neurosci.* **8,** 437–445.

Groenewegen, H. J. and Berendse, H. W. (1990) Connections of the subthalamic nucleus with ventral striatopallidal parts of the basal ganglia in the rat. *J. Comp. Neurol.* **294,** 607–622.

Hayashi, Y., Tanabe, Y., Aramori, I., Masu, M., Shimamoto, K., Ohfune, Y., and Nakanishi, S. (1992) Agonist analysis of 2-(carboxycyclopropyl)glycine isomers for cloned metabotropic glutamate receptor subtypes expressed in Chinese hamster ovary cells. *Br. J. Pharmacol.* **107,** 539–543.

Houamed, K. M., Kuijper, J. L., Gilbert, T. L., Haldeman, B. A., O'Hara, P. J., Mulvihill, E. R., Almers, W., and Hagen, F. S. (1991) Cloning, expression, and gene structure of a G protein-coupled glutamate receptor from rat brain. *Science* **252,** 1318–1321.

Hwang, P. M., Bredt, D. S., and Snyder, S. H. (1990) Autoradiographic imaging of phosphoinositide turnover in the brain. *Science* **249,** 802–804.

Ito, M., Sakurai, M., and Tongroach, P. (1982) Climbing fibre induced depression of both mossy fiber responsiveness and glutamate sensitivity of cerebellar Purkinje cells. *J. Physiol.* **324,** 113–134.

Kano, M. and Kato, M. (1987) Quisqualate receptors are specifically involved in cerebellar synaptic plasticity. *Nature* **325,** 276–279.

Kim, J. S., Hassler, R., Haug, P., and Paik, K. S. (1977) Effect of frontal cortex ablation on striatal glutamic acid level in the rat. *Brain Res.* **132,** 370–374.

Kita, H. and Kitai, S. T. (1987) Efferent projections of the subthalamic nucleus in the rat: light and electron microscopic analysis with the PHA-L method. *J. Comp. Neurol.* **260,** 435–452.

Lee, H. J., Rye, D. B., Hallanger, A. E., Levey, A. I., and Wainer, B. H. (1988) Cholinergic vs. noncholinergic efferents from the mesopontine tegmentum to the extrapyramidal motor system nuclei. *J. Comp. Neurol.* **275,** 469–492.

Linden, D. J., Dickinson, M. H., Smeyne, M., and Connor, J. A. (1991) A long-term depression of AMPA currents in cultured cerebellar Purkinje neurons. *Neuron* **7,** 81–89.

Martin, L. J., Blackstone, C. D., Huganir, R. L., and Price, D. L. (1992) Cellular localization of a metabotropic glutamate receptor in rat brain. *Neuron* **9,** 259–270.

Masu, M., Tanabe, Y., Tsuchida, K., Shigemoto, R., and Nakanishi, S. (1991) Sequence and expression of a metabotropic glutamate receptor. *Nature* **349,** 760–765.

McDonald, J. W. and Schoepp, D. D. (1992) The metabotropic excitatory amino acid receptor agonist 1S,3R-ACPD selectively potentiates *N*-methyl-D-aspartate-induced brain injury. *Eur. J. Pharmacol.* **215,** 353,334.

McGeer, P. L., McGeer, E. G., Scherer, U., and Singh, K. (1977) A glutamatergic corticostriatal path? *Brain Res.* **128,** 369–373.

Minakami, R., Katsuki, F., and Sugiyama, H. (1993) A variant of metabotropic glutamate receptor subtype 5: an evolutionarily conserved insertion with no termination codon. *Biochem. Biophys. Res. Commun.* **194,** 622–627.

Nakajima, Y., Iwakabe, H., Akazawa, C., Nawa, H., Shigemoto, R., and Mizuno, N. (1993) Molecular characterization of a novel retinal metabotropic glutamate receptor mGluR6 with a high agonist selectivity for L-2-amino-4-phosphonobutyrate. *J. Biol. Chem.* **268**, 11,868–11,873.

Nakanishi, H., Kita, H., and Kitai, S. T. (1987) Intracellular study of rat substantia nigra pars reticulata neurons in an *in vitro* slice preparation: electrical membrane properties and response characteristics to subthalamic stimulation. *Brain Res.* **437**, 45–55.

Nakanishi, S. (1992) Molecular diversity of glutamate receptors and implications for brain function. *Science* **258**, 597–603.

Nicoletti, F., Iadarola, M. J., Wroblewski, J. T., and Costa, E. (1986) Excitatory amino acid recognition sites coupled with inositol phospholipid metabolism: developmental changes and interaction with α_1-adrenoreceptors. *Proc. Natl. Acad. Sci. USA* **83**, 1931–1935.

Nicoletti, F., Wroblewski, J. T., Alho, H., Eva, C., Fadda, E., and Costa, E. (1987) Lesions of putative glutamatergic pathways potentiate the increase of inositol phospholipid hydrolysis elicited by excitatory amino acids. *Brain Res.* **436**, 103–112.

Ohishi, H., Shigemoto, R., Nakanishi, S., and Mizuno, N. (1993a) Distribution of the messenger RNA for a metabotropic glutamate receptor, mGluR2, in the central nervous system of the rat. *Neurosci.* **53**, 1009–1018.

Ohishi, H., Shigemoto, R., Nakanishi, S., and Mizuno, N. (1993b) Distribution of the mRNA for a metabotropic glutamate receptor (mGluR3) in the brain: an in situ hybridization study. *J. Comp. Neurol.* **335**, 252–266.

Pickering, D. S., Thomsen, C., Suzdak, P. D., Fletcher, E. J., Robitaille, R., Salter, M. W., MacDonald, J. F., Huang, X.-P., and Hampson, D. R. (1993) A comparison of two alternatively spliced forms of a metabotropic glutamate receptor coupled to phosphoinositide turnover. *J. Neurochem.* **61**, 85–92.

Pin, J. P., Waeber, C., Prezeau, L., Bockaert, J., and Heinemann, S. F. (1992) Alternative splicing generates metabotropic glutamate receptors inducing different patterns of calcium release in *Xenopus* oocytes. *Proc. Natl. Acad. Sci. USA* **89**, 10,331–10,335.

Price, R. H., Hollingsworth, Z., Young, A. B., and Penney, J. B. (1993) Excitatory amino acid receptor regulation after subthalamic nucleus lesions in the rat. *Brain Res.* **602**, 157–160.

Robledo, P. and Feger, J. (1990) Excitatory influence of rat subthalamic nucleus to substantia nigra pars reticulata and the pallidal complex: electrophysiological data. *Brain Res.* **518**, 47–54.

Sacaan, A. I. and Schoepp, D. D. (1992) Activation of hippocampal metabotropic excitatory amino acid receptors leads to seizures and neuronal damage. *Neurosci. Lett.* **139**, 77–82.

Sacaan, A. I., Monn, J. A., and Schoepp, D. D. (1991) Intrastriatal injection of a selective metabotropic excitatory amino acid receptor agonist induces contralateral turning in the rat. *J. Pharmacol. Exp. Ther.* **259**, 1366–1370.

Sacaan, A. I., Bymaster, F. P., and Schoepp, D. D. (1992) Metabotropic glutamate receptor activation produces extrapyramidal motor system activation that is mediated by dopamine. *J. Neurochem.* **59**, 245–251.

Seren, M. S., Aldinio, C., Zanoni, R., Leon, A., and Nicoletti, F. (1989) Stimulation of inositol phospholipid hydrolysis by excitatory amino acids is enhanced in brain slices from vulnerable regions after transient global ischemia. *J. Neurochem.* **53**, 1700–1705.

Shigemoto, R., Nakanishi, S., and Mizuno, N. (1992) Distribution of the mRNA for a metabotropic glutamate receptor (mGluR1) in the central nervous system: an *in situ* hybridization study in adult and developing rat. *J. Comp. Neurol.* **322**, 121–135.

Sladeczak, F., Pin, J.-P., Récasens, M., Bockaert, J., and Weiss, S. (1985) Glutamate stimulates inositol phosphate formation in striatal neurons. *Nature* **317**, 717–719.

Smith, Y. and Parent A. (1988) Neurons of the subthalamic nucleus in primates display glutamate but not GABA immunoreactivity. *Brain Res.* **453**, 353–356.

Standaert, D. G., Testa, C. M., Penney, J. B., and Young, A. B. (1993) Alternatively spliced isoforms of the NMDAR1 glutamate receptor: differential expression in the basal ganglia of the rat. *Neurosci. Lett.* **152**, 161–164.

Sugiyama, H., Ito, I., and Hirono C. (1987) A new type of glutamate receptor linked to inositol phospholipid metabolism. *Nature* **325**, 531–533.

Tanabe, Y., Masu, M., Ishii, T., Shigemoto, R., and Nakanishi, S. (1992) A family of metabotropic glutamate receptors. *Neuron* **8**, 169–179.

Tanabe, Y., Nomura, A., Masu, M., Shigemoto, R., Mizuno, N., and Nakanishi S. (1993) Signal transduction, pharmacologic properties, and expression patterns of two rat metabotropic glutamate receptors, mGluR3 and mGluR4. *J. Neurosci.* **13**, 1372–1378.

Testa, C. M., Standaert, D. G., Young, A. B., and Penney, J. B. (1994) Metabotropic glutamate receptor expression in the basal ganglia of the rat. *J. Neurosci.* in press.

Winder, D. G. and Conn, P. J. (1993) Activation of metabotropic glutamate receptors increases cAMP accumulation in hippocampus by potentiating responses to endogenous adenosine. *J. Neurosci.* **13**, 38–44.

Worley, P. F., Baraban, J. M., and Snyder, S. H. (1989) Inositol 1,4,5-trisphosphate receptor binding: autoradiographic localization in rat brain. *J. Neurosci.* **9**, 339–346.

Worley, P. F., Baraban, J. M., De Souza, E. B., and Snyder, S. H. (1986a) Mapping second messenger systems in the brain: differential localizations of adenylyl cyclase and protein kinase C. *Proc. Natl. Acad. Sci. USA* **83**, 4053–4057.

Worley, P. F., Baraban, J. M., and Snyder, S. H. (1986b) Heterogeneous localization of protein kinase C in rat brain: autoradiographic analysis of phobol ester receptor binding. *J. Neurosci.* **6**, 199–207.

Young, A. B., Bromberg, M. B., and Penney, J. B. (1981) Decreased glutamate uptake in subcortical areas deafferented by sensorimotor cortical ablation in the cat. *J. Neurosci.* **1**, 241–249.

Young, A. B. and Fagg, G. E. (1990) Excitatory amino acid receptors in the brain: membrane binding and receptor autoradiographic approaches. *Trends Pharmacol. Sci.* **11**, 126–133.

CHAPTER 5

Modulation of Ionic Currents by Metabotropic Glutamate Receptors in the CNS

Urs Gerber and Beat H. Gähwiler

1. Introduction

Stimulation of neuronal metabotropic glutamate receptors (mGluRs) activates G-proteins initiating a multitude of intracellular processes. Electrophysiological observations indicate that ion channels are among the substrates for the intracellular messengers dispatched by activated mGluRs, thus modulating the electrical behavior of neurons owing to their coupling to numerous membrane ionic conductances. In some of the first experiments to demonstrate metabotropic effects of glutamate, electrophysiological techniques were employed using transfected oocytes to assay for changes in second-messenger levels in response to glutamate (Sugiyama et al., 1987, 1989). It has only recently been shown that activation of mGluRs can also modify the firing properties of neurons (Stratton et al., 1989, 1990; Baskys et al., 1990; Charpak et al., 1990; Desai and Conn, 1991).

The Metabotropic Glutamate Receptors Eds.: P. J. Conn and J. Patel
© 1994 Humana Press Inc., Totowa, NJ

Results will be presented in this chapter from studies that have characterized the receptors, the signal transduction mechanisms, and the ionic conductances mediating electrophysiological responses to mGluR stimulation in the vertebrate central nervous system. Most of the observations to be described are from experiments in which indirect actions resulting from presynaptic effects could be ruled out, i.e., recordings were obtained from single isolated neurons, or responses were shown to persist under conditions where propagated electrical activity was blocked with tetrodotoxin, or synaptic transmission was disrupted with low-calcium/high magnesium-containing perfusion solutions. Some of the material that will be discussed here has been covered in previous reviews (*see* Schoepp et al., 1990; Conn and Desai, 1991; Miller, 1991; Anwyl, 1992; Baskys, 1992; Nakanishi, 1992; Zorumski and Thio, 1992; Schoepp and Conn, 1993). In this chapter, our aim is to update the field, focusing on the role of postsynaptic mechanisms in the modulation of neuronal activity.

2. Modulation of Potassium Currents

Membrane potassium currents are crucial in maintaining the level of excitability in the nervous system within a tractable range. Potassium currents set the resting membrane potential of neurons, promote repolarization of action potentials, and modulate the rate of repetitive spike discharge. By regulating the gating of potassium channels, mGluRs contribute to the control of neuronal function.

2.1. Calcium-Dependent Potassium Currents

The first evidence indicating that activation of mGluRs can modify neuronal membrane properties came from experiments that showed that exposing hippocampal pyramidal cells to glutamate or related agonists inhibits accommodation of cell firing and diminishes the slow calcium-dependent after-hyperpolarization following a train of action potentials (Stratton et al., 1989, 1990; Baskys et al., 1990; Charpak et al., 1990; Desai and Conn, 1991). These excitatory effects are not compatible with ionotropic responses to glutamate and are maintained in the presence of ionotropic receptor antagonists (Charpak et al., 1990; Charpak and Gähwiler, 1991). The sustained firing response of neurons observed in the presence of glutamatergic agonists results from a reduction in a slow, calcium-dependent potassium current referred to as I_{AHP} (Baskys et al., 1990; Charpak et al., 1990). I_{AHP} is generated when the intracellular calcium concentration rises because of influx through voltage-gated calcium channels, leading to activation of low-conductance "SK" potassium channels (Lancaster and Adams, 1986; Lancaster et al., 1991). Inhibition of I_{AHP} on stimulation of mGluRs could result from a reduction in either calcium influx or potassium efflux. Microfluorometric (Fura-2) measurements combined with voltage-clamp recording have shown that low concentrations of metabotropic agonists (quisqualate ≤ 1 μM) can inhibit potassium currents without affecting calcium influx (Charpak et al., 1990). A similar effect on potassium current, independent of changes in intracellular calcium, has also been reported for neurons in the nucleus of the solitary tract (Glaum and Miller, 1992). These studies show that low concentrations of agonists activate mGluRs, which selectively couple to a potassium conductance. Furthermore, the results imply that the intracellular transduction mechanism mediating the reduction in I_{AHP} does not directly depend on calcium. As will be discussed below, higher concentrations of mGluR agonists do cause a reduction in calcium current.

Activation of mGluRs is also reported to increase a calcium-dependent potassium current in hippocampal pyramidal cells (Akaike et al., 1991). To date, this effect has only been observed in acutely dissociated hippocampal pyramidal cells from young rats, suggesting that in other preparations this action is occluded by other responses. This enhanced current in response to glutamatergic agonists may be the result of activation of a large conductance calcium-dependent potassium channel as has been reported in cerebellar granule cells (Fagni et al., 1991). However, it appears that mGluRs do not gate these potassium channels directly, but rather augment the levels of intracellular calcium through IP_3-mediated release and increased calcium channel activity (Fagni et al., 1991; Bossu et al., 1992).

2.2. Voltage-Dependent Potassium Current

In addition to promoting cell firing, stimulation of mGluRs is also found to depolarize hippocampal pyramidal cells and interneurons (Stratton et al., 1989, 1990; Charpak et al., 1990; Desai and Conn, 1991; Hu and Storm, 1992; Miles and Poncer, 1993). The depolarization, associated with a decrease in membrane resistance, is the result of a reduction in potassium conductance, since the response depends on extracellular potassium concentration in an Nernstian manner (Charpak et al., 1990) and is not affected by changing the chloride gradient (Stratton et al., 1990; Charpak et al., 1990). The possibility that the depolarization might be the result of the blockade of a tonically active component of I_{AHP} when mGluRs are activated can be ruled out. When I_{AHP} was blocked with intracellular cyclic AMP (Madison and Nicoll, 1986), depolarization or inward current in response to agonists is not reduced (Stratton et al., 1990; Guérineau et al., 1994). Other maneuvers to abolish calcium-dependent I_{AHP}, such as filling cells with the calcium chelator BAPTA, preventing calcium influx by removing extracellular calcium, or applying cadmium, also fail to reduce mGluR-mediated inward current significantly (Charpak et al., 1990).

With I_{AHP} blocked, activation of mGluRs can be shown to decrease the time- and voltage-dependent potassium current known as I_M (Charpak et al., 1990). That the current was indeed I_M could be established by demonstrating that application of cholinergic, muscarinic agonists (Brown and Adams, 1980) occluded the action of mGluR agonists. I_M is activated at membrane potentials slightly depolarized to resting potential (~ -60 mV) in hippocampal neurons (Halliwell and Adams, 1982). Reduction of I_M thus contributes to the depolarization in response to mGluR stimulation and would be expected to potentiate concurrent postsynaptic potentials.

2.3. Voltage-Independent Potassium Current

Glutamatergic metabotropic agonists have been shown to depolarize thalamic relay cells (McCormick and von Krosigk, 1992), which do not exhibit significant I_M and I_{AHP} (McCormick, 1991). The depolarization is associated with a decrease in a potassium conductance that, according to its current–voltage relationship, is not voltage-dependent. This is consistent with a reduction of a resting "leak" conductance ($I_{K, leak}$; McCormick and von Krosigk, 1992). It was of interest to establish whether inhibition of $I_{K, leak}$ is also involved in metabotropic responses in other neurons. To answer this question, Guérineau et al. (1994) recorded from CA3 hippocampal pyramidal cells using the whole-cell, patch-clamp technique under conditions where I_M and I_{AHP} were not active: dialysis with cyclic AMP to inhibit I_{AHP} and voltage-clamped at a membrane potential negative to the activation range for I_M. In this situation, stimulation of mGluRs induced an inward current associated with a decrease in membrane conductance, which reversed close to the equilibrium potential for potassium and varied linearly with membrane potential. These experiments established that reduction of a potassium current corresponding to $I_{K, leak}$ also contributes to the inward current observed with mGluR stimulation in hippocampal cells.

In hippocampal pyramidal cells, mGluR agonists cause a broadening of sodium-dependent action potentials owing to a prolongation of the repolarization phase (Hu and Storm, 1991). This suggests that, here again, mGluRs reduce an as yet unidentified potassium conductance.

In Müller glial cells of the retina, mGluRs also couple to a potassium conductance corresponding to a membrane leak conductance (Schwartz, 1993). Although it is not known whether the channels mediating this leak conductance are the same as those in thalamic and hippocampal cells, it is clear that different intracellular transduction pathways are involved (*see* Section 2.4.).

2.4. Transduction Mechanisms

2.4.1. G-Proteins

The molecular structure of mGluRs, which is characterized by seven putative transmembrane domains (*see* Chapter 7), indicates that an ion-conducting pore is not formed. An intracellular transduction mechanism is necessary for these receptors to modify the activity of ion channels. As is the case for other metabotropic responses, it has been shown that activation of membrane-bound heterotrimeric G-proteins is involved in the modulation of potassium conductances. The binding of agonist to a metabotropic receptor promotes dissociation of GDP from the inactive α,β,γ trimer, allowing GTP to bind, which then activates and liberates the $G\alpha$ and the $G\beta\gamma$ subunits for subsequent interaction with appropriate effectors (Gilman, 1987). Using nonhydrolyzable analogs of GTP, Guérineau et al. (1994) have demonstrated that CA3 pyramidal cells dialyzed with GDPβS, which prevents G-protein activation, no longer show a reduction in $I_{K, leak}$ when exposed to metabotropic agonists, whereas cells dialyzed with GTPγS, which irreversibly activates G-proteins, exhibit enhanced and sustained responses. Using the same approach, Schwartz (1993) demonstrated that in Müller glial cells, the slow phase of the reduction in potassium conductance is greatly diminished when GDPβS is introduced into cells. As shown in Fig. 1, in CA3 cells mGluRs mediating the reduction in I_{AHP} are also coupled to GDPβ-sensitive G-proteins. Control responses were obtained using *trans*-1-amino-cyclopentyl-1,3-dicarboxylic acid (trans ACPD), a selective agonist of mGluRs, and adenosine, which is known to act on G-protein-coupled receptors (Trussell and Jackson, 1987). After allowing GDPbS (as a lithium salt) to diffuse into the cell, subsequent application of agonists induced greatly reduced responses in the same cell (Fig. 1B). In control experiments using equivalent concentrations of lithium alone, no effects on responses were noted. These observations have also been reported in a preliminary study on hippocampal dentate granule cells (Baskys et al., 1990).

In experiments aimed at identifying the subtype of G-protein linking mGluRs to potassium conductances in the hippocampus, responses were found not to be affected by pretreatment of cells with pertussis toxin (Stratton et al., 1990; Gerber et al., 1992), which inactivates G_o-α and G_i-α subunits by ADP-dependent ribosylation (Katada and Ui, 1982). A member of the new group of G-proteins, the G_q subfamily, represents a likely candidate for these responses (Bockaert, 1991; Sternweis and Smrcka, 1992). Final identification of the G-protein awaits experiments in which subunit-specific, affinity-purified antibodies suitable for immunoprecipitation are introduced into the cells in order to block responses. In the case of Müller glial cells, the slow phase of glutamate-induced reduction of potassium current is sensitive to pertussis toxin treatment, establishing a role for a $G_{i/o}$-type G-protein in these responses (Schwartz, 1993).

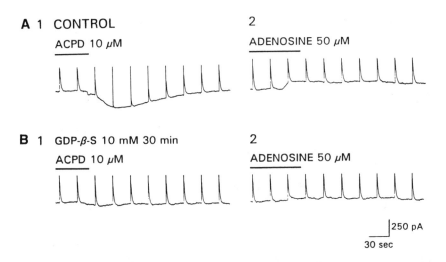

Fig. 1. Response to ACPD of a hippocampal CA3 pyramidal cell dialyzed with GDPβS. Chart records show membrane current in a cell voltage-clamped at −55 mV in the presence of tetrodotoxin (0.5 μ*M*). Brief depolarizing voltage steps (50 ms, 40 mV) were applied every 25 s to evoke outward tail currents (I_{AHP}). **(A)** Within 5 min after introduction of a sharp microelectrode containing GDPβS, a normal control response to ACPD and adenosine was obtained. ACPD induced an inward current and reduced the amplitude of I_{AHP} (A1), and adenosine induced an outward current (A2). **(B)** After 30 min, GDPβS has diffused into the cell causing a profound reduction in responses to ACPD (B1) and adenosine (B2).

2.4.2. Second Messengers

What is the target of activated G-protein subunits? It has become apparent that enzymes and ion channels share both structural and functional characteristics (Jan and Jan, 1992). There are thus two scenarios whereby G-proteins could regulate ion channel activity: a direct effect of G-proteins on channels (Yatani et al., 1987; Kubo et al., 1993) or an activation by G-proteins of enzymes that then regulate the gating properties of channels.

Numerous studies have demonstrated that phosphorylation, the primary regulatory mechanism of enzyme activity, is important for the regulation of gating properties of most ion channels (Kaczmarek and Levitan, 1987). Previous work had indeed shown that activation of protein kinase C (PKC) with phorbol esters has a depressant effect on I_{AHP} in hippocampal pyramidal cells (Baraban et al., 1985; Malenka et al., 1986) indistinguishable from that induced by stimulation of mGluRs. A study was therefore conducted to examine whether mGluRs reduce I_{AHP} through activation of PKC (Gerber et al., 1992). In the presence of broad-spectrum kinase inhibitors, such as staurosporine or H7 (Rüegg and Burgess, 1989), the effect of phorbol esters was totally blocked, indicating that activation of PKC was prevented (Fig. 2B, bottom trace). Under these conditions, however, the inhibition of I_{AHP} by mGluR activation remains unchanged (Fig. 2). Further experiments demonstrated that neither PKC nor protein kinase A (PKA) activation is required for this response (Gerber et al., 1992). A recent study has confirmed and extended these findings by showing that calcium calmodulin kinase II is also not involved in the reduction of I_{AHP} following stimulation of mGluRs (Müller et al., 1992). These negative results suggest that G-proteins may directly couple mGluRs to potassium channels in hippocampal neurons.

In contrast, $G_{i/o}$ type G-proteins that couple to mGluRs in Müller glial cells appear to interact with adenylyl cyclase, leading to the activation of PKA (Schwartz, 1993). Since only the G_s-type α subunit is known to activate adenylyl cyclase, whereas the $G_{i/o}$-type α subunits inhibit this enzyme, Schwartz (1993) has suggested that these glial G-proteins are contributing $\beta\gamma$ subunits, which can significantly potentiate the activity of adenylyl cyclase (Tang and Gilman, 1991; Federman et al., 1992). Direct application of the relevant G-protein subunits and second-messenger candidates to membrane patches displaying single potassium channel activity will be required to verify the transduction mechanisms governing potassium conductance.

2.5. Agonists and Antagonists

In order to characterize the mGluRs-mediating electrophysiological responses, neurophysiologists are dependent on the availability of the appropriate pharmacological implements. A number of endogenous amino acids, including glutamate, aspartate, cysteine, and homocysteine, have been shown to activate mGluRs coupled to potassium conductances, although not selectively (Charpak et al., 1990). The observation that *trans*-ACPD selectively activates

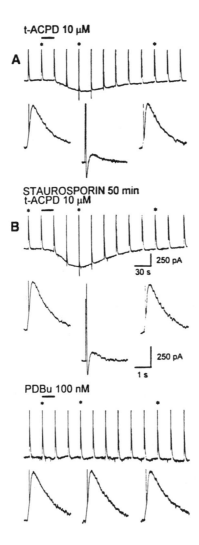

Fig. 2. Effects of ACPD on membrane currents in a CA3 pyramidal cell before and after application of staurosporin. The experimental conditions are the same as in the previous figure. Asterisks indicate tail currents illustrated at an expanded time scale below each trace. **(A)** ACPD induced an inward current and reduced the amplitude of I_{AHP} under control conditions. **(B)** Following exposure of the slice culture to staurosporin, the response to ACPD was essentially unchanged (middle panel). Under these conditions, phorbol esters were without effect in the same cell, indicating that activation of PKC was prevented (bottom panel).

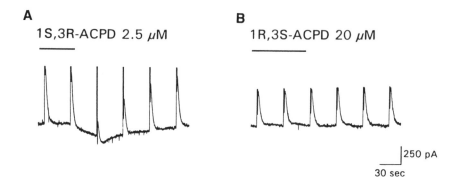

Fig. 3. Stereospecificity of 1S,3R-ACPD at mGluRs mediating reduction in
K⁺ currents. The experimental conditions are the same as in previous figures.
(A) 1S,3R-ACPD induced an inward current and reduced the amplitude of
I_{AHP}. **(B)** No effect was observed when the enantiomer 1R,3S-ACPD was
applied to the same cell, despite the application of a higher concentration.

mGluRs has greatly facilitated the investigation of these responses
(Palmer et al., 1989; Manzoni et al., 1990; Desai and Conn, 1990).
1S,3R-ACPD appears to be the most potent enantiomer of ACPD at
most metabotropic receptors (Irving et al., 1990; Schoepp and Conn,
1993), including those that induce reduction of potassium conductances (Desai et al., 1992; Guérineau et al., 1994; Fig. 3).

Progress in the discovery and cloning of novel mGluR subtypes
has advanced more rapidly than the development of subtype-specific
pharmacological agents. Agonist potency profiles have been determined for the cloned and expressed mGluR subtypes (Nakanishi,
1992) that permit tentative identification of receptors. With this
approach, the mGluRs that modulate potassium conductance have
been shown to correspond in certain cases to the type 1 or type 5 class
(Desai et al., 1992; Schwartz, 1993).

The elucidation of the physiological roles for mGluRs has been
delayed by the lack of potent and selective antagonists. A number of
proposed antagonists, notably aminophosphonopropionate (AP3),
were found to be ineffective in blocking electrophysiological
responses mediated by mGluRs (*see* Gerber et al., 1993).

Recently a selective and competitive antagonist has become available: (RS)-α-methyl-4-carboxyphenylglycine (MCPG) (Bashir et al. 1993; Eaton et al., 1993). Using this antagonist, Bashir and associates (1993) have provided evidence that activation of mGluRs is required for the induction of long-term potentiation. MCPG is also an effective antagonist at mGluRs that regulate several membrane potassium conductances in hippocampal pyramidal cells (Gerber et al., 1993; Guérineau et al., 1994). This has permitted the demonstration that postsynaptic mGluRs can be activated by synaptically released excitatory amino acids, evoking a long-lasting inward current and reducing the calcium-dependent potassium current underlying accommodation of cell firing, as previously proposed (Charpak and Gähwiler, 1991). Slow synaptic responses were evoked in CA3 cells by stimulating mossy fibers during blockade of NMDA, AMPA/ kainate, and GABA receptors (Fig. 4A; Charpak and Gähwiler, 1991). MCPG reversibly abolishes this synaptically activated inward current and the synaptic inhibition of evoked I_{AHP} (Fig. 4B; Gerber et al., 1993). These results show that mGluRs, known to gate potassium conductances in the hippocampus, can contribute to synaptic responses in CA3 cells evoked by stimulation of mossy fibers. Thus, synaptic modulation of potassium conductances by glutamate may underlie long-term increases in neuronal excitability.

2.6. Interactions with Other Receptors

The use of selective agonists was essential for the initial characterization of responses mediated by mGluRs. Under physiological conditions, however, it is unlikely that mGluRs will be selectively activated. This raises the question of whether functional interactions occur when ionotropic and metabotropic receptors are simultaneously activated by synaptically released glutamate. Indeed, several laboratories have reported that activation of mGluRs can potentiate ionotropic glutamate responses in various preparations (Aniksztejn et al., 1992; Bleakman et al., 1992; Cerne and Randic, 1992; Kinney and Slater, 1992; Collins, 1993; Harvey and Collingridge, 1993). Preliminary work has shown that the converse, namely,

A

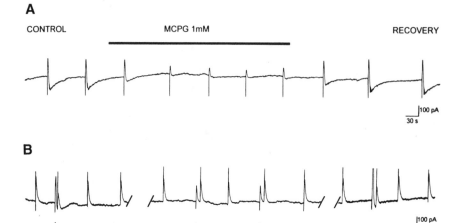

Fig. 4. Actions of the metabotropic glutamatergic antagonist, MCPG, on slow evoked synaptic responses. Mossy fibers were electrically stimulated (5 pulses, 100 Hz, 100 μs, 50 μA) to evoke a slow synaptic current in a voltage-clamped CA3 pyramidal cell under conditions of GABA and ionotropic glutamate receptor blockade with bicuculline (20 μM), D-APV (40 μM), and CNQX (40 μM). **(A)** The continuous chart recording depicts the rapid and reversible antagonistic action of MCPG on slow synaptic responses. **(B)** In this cell, I_{AHP} was periodically evoked with brief depolarizing voltage steps from -55 to -15 mV, 50-ms duration, every 30 s. Arrows indicate when mossy fibers were stimulated, evoking an inward current and inhibition of I_{AHP}. MCPG reversibly attenuated the reduction in I_{AHP} elicited by synaptic stimulation.

a potentiation of ACPD-induced current after NMDA receptor activation, is also possible (Lüthi et al., 1993). These interactions offer novel postsynaptic mechanisms that may contribute to the activity-dependent changes in synaptic efficacy observed at glutamatergic synapses (*see* Chapter 7).

3. Modulation of Calcium Currents

Calcium channel gating represents an important mechanism for regulation of cytoplasmic calcium concentration and the many calcium-dependent intracellular processes. Although studies generally focus on somatic calcium channels for technical reasons, mGluRs and the calcium channels that they govern may also be located on presynaptic terminals where they could modulate transmitter release, as discussed in Chapter 6.

3.1. L-Type Calcium Conductance

The first description of the effects of glutamate on calcium conductance demonstrated that activation of a G-protein-coupled receptor causes a reduction in high-threshold calcium current in CA1 pyramidal cells of the hippocampus (Lester and Jahr, 1990). In this study, the subtype of calcium channels mediating the high-threshold current was not determined, but subsequent work has revealed that both L- and N-type currents can be reduced by glutamate in hippocampal cells (Sahara and Westbrook, 1993). Reduction of L-type current is characterized by relatively slow kinetics, and required raising the intracellular calcium concentration to 100 nM.

Neocortical pyramidal neurons exhibit T-, L-, and N-type calcium channels. In these cells, only the dihydropyridine-sensitive L-type current is reduced with stimulation of mGluRs (Sayer et al., 1992). Based on the agonist potency profile, the receptor mediating this effect corresponds most closely to the mGluR1 or mGluR5. The response is not modified under conditions where intracellular calcium stores were depleted or chelated. However, this process shows some calcium dependency, since responses were reduced when barium was substituted for calcium (Sayer et al., 1992). A single channel study of L-type current in cerebellar granule cells has shown that mGluR agonists reduces channel activity via a pertussis toxin-sensitive G-protein in these cells (Chavis et al., 1993). In view of the sensitivity of this receptor to L-2-carboxycyclopropyl-glycine (L-CCGI), this receptor was provisionally identified as a mGluR2 (*see* Chapter 1). The transduction mechanism, which does not involve kinases or phosphatases, remains unclear.

Some of the cloned mGluRs are potently activated by 2-amino-4-phosphonobutanoate (L-AP4) (*see* Chapter 1). Furthermore, L-AP4 acts on a G-protein-coupled glutamate receptor to inhibit high-threshold calcium currents in olfactory bulb neurons. This pertussis toxin-sensitive response persists when known second-messenger pathways are disrupted, leading to the suggestion that activated G-proteins might directly interact with calcium channels (Trombley and Westbrook, 1992).

To date, one example of enhancement of L-type channel activity by mGluRs has been reported. In cerebellar granule cells, application of ACPD increased calcium-channel activity, an effect that became more apparent in pertussis toxin-treated neurons in which the previously described inhibitory action on channel activity was suppressed (Bossu et al., 1992, 1993; Zegarra-Moran and Moran, 1993). It thus appears that the one mGluR (perhaps mGluR 1 or 5) is positively coupled to L-type calcium channels in cerebellar granule cells, whereas an mGluR2-like receptor is negatively coupled (Fagni et al., 1993).

3.2. N-Type Calcium Conductance

Activation of mGluRs can also induce an inhibition of N-type calcium current, based on its sensitivity to ω-conotoxin GVIA in hippocampal pyramidal cells (Swartz and Bean, 1992; Sahara and Westbrook, 1993). According to the rank order of agonist potency, this receptor may be of the mGluR1 or mGluR5 subtype. It appears that the G-protein involved in this response is not easily inactivated by pertussis toxin and may therefore be only partially sensitive to this treatment, or that more than one type of G-protein may be involved (Lester and Jahr, 1990; Swartz and Bean, 1992; Sahara and Westbrook, 1993). Application of 1S,3R-ACPD outside the patch, when recording in the cell-attached mode, does not inhibit the activity of calcium channels. However, calcium-channel currents are suppressed when outside-out membrane patches are exposed to 1S,3R-ACPD. This provides strong evidence for a membrane-delimited pathway involving a direct action of G-proteins on calcium channels (Swartz and Bean, 1992). The demonstration of calcium-channel gating by mGluRs represents a further mechanism that may contribute to the phenomena of synaptic plasticity, a hallmark of glutamatergic synapses.

4. Modulation of Cationic Currents

4.1. After-Depolarization Current

The induction or potentiation of a slow poststimulus after-depolarization (ADP) by mGluR agonists has been observed in neurons of the dorsolateral septal nucleus (Zheng and Gallagher, 1991, 1992a,b), the olfactory cortex (Constanti and Libri, 1992), the neocortex (Greene et al., 1992), and the hippocampus (Caeser et al., 1993). This response, associated with a decrease in membrane resistance and dependent on extracellular calcium, appears to be equivalent to the previously described ADP observed with application of cholinergic muscarinic agonists to neurons (*see* Colino and Halliwell, 1993). In both cases, a slow calcium-dependent cationic current was induced that overrides the normally occurring poststimulus after-hyperpolarization (Zheng and Gallagher, 1992; Caeser et al., 1993). This current promotes a shift of neurons into a burst-firing mode that may have implications both for synaptic plasticity and pathological events, such as epilepsy.

4.2. Inward/Outward Currents in Purkinje Cells

Cerebellar Purkinje cells respond to mGluR agonists with a depolarization or inward current as is seen in other neurons (Ito and Karachot; 1990; Crépel et al., 1991; Vranesic et al., 1991; East and Garthwaite, 1992; Glaum et al., 1992; Kinney and Slater, 1992; Staub et al., 1992; Yool et al., 1992). However, this response, which is associated with a slight increase in conductance and a reversal potential close to +20 mV, is not mediated by a reduction in potassium conductance. The properties of this inward current are consistent with the activation of calcium-dependent nonspecific cation channels (Staub et al., 1992). Another plausible mechanism could be the activation of an electrogenic sodium/calcium exchanger in response to increased intracellular levels of calcium induced by mGluR stimulation (Staub et al., 1992; Glaum et al., 1992).

A number of recent studies have shown that an outward current associated with a decrease in membrane conductance followed the inward current induced by activation of mGluRs in Purkinje cells (Inoue et al., 1992; Takagi et al., 1992; Vranesic et al., 1993). This response was not associated with changes in intracellular calcium and exhibited a reversal potential close to 0 mV, indicating the reduction of a tonically active resting conductance, for example, calcium conductance (Inoue et al., 1992; Vranesic et al., 1993). The observation that the current was not observed in immature Purkinje cells, lacking well-developed dendrites, suggests the possibility of a dendritic localization of the receptors or ion channels involved (Takagi et al., 1992).

These results indicate that Purkinje cells, which exhibit high densities of mGluRs (Masu et al., 1991), employ alternative ionic mechanisms in the production of electrophysiological signals. Although these receptors have been implicated in the phenomenon of cerebellar LTD (Ito and Karachot, 1990), their precise role in these plastic changes remains unclear at present (Glaum et al., 1992).

5. Conclusions and Physiological Implications

The widespread distribution of mGluRs and the prevalence of glutamate and related endogenous agonists throughout the nervous system are indicative of important physiological functions that are now beginning to be elucidated. We have presented evidence for the modulation by mGluRs of a wide variety of ionic conductances in diverse neuronal cell types and have reviewed what is known about the mechanisms underlying the regulation of channel activity. The demonstration that excitatory amino acids are not only responsible for fast neurotransmission, but also exert slower actions mediated by various potassium, calcium, and cationic currents suggests various new roles for these neurotransmitters in the control of nervous system function. By reducing potassium conductances, mGluRs can potentially modulate background tone in key nuclei and thus contribute to the control of behavioral state (McCormick and von Krosigk 1992; Wang and McCormick 1993). Effects on calcium channels allow the modulation of synaptic transmission and intracellular calcium-dependent processes. As our understanding of intracellular transduction associated with mGluRs improves, we shall gain new insights into the mechanisms of interaction with other neurotransmitters systems.

Acknowledgments

The authors wish to thank Dr. Scott Thompson for reading the manuscript. This work was supported by the Swiss National Science Foundation (grant 31-35976.92) and the Prof. Dr. Max Cloëtta Foundation.

References

Akaike, N., Shirasaki, T., and Harata, N. (1991) Metabotropic glutamate receptors activates the potassium conductance in freshly dissociated hippocampal CA1 neurons. *Soc. Neurosci.* **17,** 255.

Aniksztejn, L., Otani, S., and Ben-ari, Y. (1992) Quisqualate metabotropic receptors modulate NMDA currents and facilitate induction of long-term potentiation through protein kinase C. *Eur. J. Neurosci.* **4,** 500–505.

Anwyl, R. (1992) Metabotropic glutamate receptors: electrophysiological properties and role in plasticity. *Rev. Neurosci.* **3,** 217–231.

Baraban, J. M., Snyder, S. H., and Alger, B. E. (1985) Protein kinase C regulates ionic conductance in hippocampal pyramidal neurons, Electrophysiological effects of phorbol esters. *Proc. Natl. Acad. Sci. USA* **82,** 2538–2542.

Bashir, Z. I., Bortolotto, Z. A., Davies, C. H., Berretta, N., Irving, A. J., Seal, A. J., Henley, J. M., Jane, D. E., Watkins, J. C., and Collingridge, G. L. (1993) Induction of LTP in the hippocampus needs synaptic activation of glutamate metabotropic receptors. *Nature* **363,** 347–350.

Baskys, A. (1992) Metabotropic receptors and "slow" excitatory actions of glutamate agonists in the hippocampus. *TINS* **15,** 92–96.

Baskys, A., Bernstein, N. K., Barolet, A. W., and Carlen, P. L. (1990) NMDA and quisqualate reduce a Ca-dependent K^+ current by a protein kinase-mediated mechanism. *Neurosci. Lett.* **112,** 76–81.

Bleakman, D., Rusin, K. I., Chard, P. S., Glaum, S. R., and Miller, R. J. (1992) Metabotropic glutamate receptors potentiate ionotropic glutamate responses in the rat dorsal horn. *Mol. Pharmacol.* **42,** 192–196.

Bockaert, J. (1991) G proteins and G-protein-coupled receptors: structure, function and interactions. *Curr. Opinion Neurobiol.* **1,** 32–42.

Bossu, J. L., Fagni, L., Nooney, J., and Feltz, A. (1992) Increased Ca channel activity due to metabotropic glutamate receptor stimulation in isolated rat cerebellar granule cells. *J. Physiol.* **459,** 250P.

Bossu, J. L., Nooney, J. M., Chavis, P., Fagni, L., Bockaert, J., and Feltz, A. (1993) Facilitatory effect on L-type Ca channels of glutamate metabotropic receptors in rodent cerebellar granule cells. *J. Neurochem.* **61,** S197B.

Brown, D. A. and Adams, P. R. (1980) Muscarinic suppression of a novel voltage-sensitive K$^+$-current in a vertebrate neuron. *Nature* **283**, 673–676.

Caeser, M., Brown, D. A., Gähwiler, B. H., and Knöpfel, T. (1993) Characterization of a calcium-dependent current generating a slow after depolarization of CA3 pyramidal cells in rat hippocampal slice cultures. *Eur. J. Neurosci.* **5**, 560–569.

Cerne, R. and Randic, M. (1992) Modulation of AMPA and NMDA responses in rat spinal dorsal horn neurons by *trans*-1-aminocyclopentane-1,3-dicarboxylic acid. *Neurosci. Lett.* **144**, 180–184.

Charpak, S. and Gähwiler, B. H. (1991) Glutamate mediates a slow synaptic response in hippocampal slice cultures. *Proc. R. Soc. Lond. (Biol.)* **243**, 221–226.

Charpak, S., Gähwiler, B. H., Do, K. Q., and Knöpfel, T. (1990) Potassium conductances in hippocampal neurons blocked by excitatory amino acid transmitters. *Nature* **347**, 765–767.

Chavis, P., Fagni, L., and Bockaert, J. (1993) Metabotropic glutamate receptors inhibit L-type calcium channels in cultured cerebellar granule cells. *J. Neurochem.* **61**, S15C.

Colino, A. and Halliwell, J. V. (1993) Carbachol potentiates Q current and activates a calcium-dependent non-specific conductance in rat hippocampus in vitro. *Eur. J. Neurosci.* **5**, 1198–1209.

Collins, G. G. S. (1993) Actions of agonists of metabotropic glutamate receptors on synaptic transmission and transmitter release in the olfactory cortex. *Br. J. Pharmacol.* **108**, 422–430.

Conn, P. J. and Desai, M. A. (1991) Pharmacology and physiology of metabotropic glutamate receptors in mammalian central nervous system. *Drug Dev. Res.* **24**, 207–229.

Constanti, A. and Libri, V. (1992) Trans-ACPD induces a slow post-stimulus inward tail current (IADP) in guinea-pig olfactory cortex neurones. *Eur. J. Pharmacol.* **214**, 105–106.

Crépel, F., Daniel, H., Hemart, N., and Jaillard, D. (1991) Effects of ACPD and AP3 on parallel-fibre-mediated EPSPs of Purkinje cells in cerebellar slices in vitro. *Exp. Brain Res.* **86**, 402–406.

Desai, M. A. and Conn, P. J. (1990) Selective activation of phosphoinositide hydrolysis by a rigid analogue of glutamate. *Neurosci. Lett.* **109**, 157–162.

Desai, M. A., Smith, T. S., and Conn, P. J. (1991) Excitatory effects of ACPD receptor activation in the hippocampus are mediated by direct effects on pyramidal cells and blockade of synaptic inhibition. *J. Neurophysiol.* **66**, 40–52.

Desai, M. A. and Conn, P. J. (1992) Multiple metabotropic glutamate receptors regulate hippocampal function. *Synapse* **12**, 206–213.

East, S. J. and Garthwaite, J. (1992) Actions of a metabotropic glutamate receptor agonist in immature and adult cerebellum. *Eur. J. Pharmacol.* **219**, 395–400.

Eaton, S. A., Jane, D. E., Jones, P. L., St., J., Porter, R. H. P., Pook, P. C. K., Sunter, D. C., Udvarhelyi, P. M., Roberts, P. J., Salt, T. E., and Watkins, J. C. (1993) Competitive antagonism at metabotropic glutamate receptors by *S*-4-carboxyphenylglycine and *RS*-α-methyl-4-carboxyphenylglycine. *Eur. J. Pharmacol.* **244**, 195–197.

Fagni, L., Bossu, J. L., and Bockaert, J. (1991) Activation of a large-conductance Ca²⁺-dependent K⁺ channel by stimulation of glutamate phosphoinositide-coupled receptors in cultured cerebellar granule cells. *Eur. J. Neurosci.* **3,** 778–789.

Fagni, L., Chavis, P., Bossu, J. L., Nooney, J. M., Feltz, A., and Bockaert, J. (1993) Control of ionic channels by metabotropic glutamate receptors. *J. Neurochem.* **61,** S198A.

Federman, A. D., Conklin, B. R., Schrader, K. A., Reed, R. R., and Bourne, H. R. (1992) Hormonal stimulation of adenylyl cyclase through G₁-protein βγ subunits. *Nature* **356,** 159–161.

Gerber, U., Lüthi, A., and Gähwiler, B. H. (1993) Inhibition of a slow synaptic response by a metabotropic glutamate receptor antagonist in hippocampal CA3 pyramidal cells. *Proc. R. Soc. Lond. (Biol.)* **254,** 169–172.

Gerber, U., Sim, J. A., and Gähwiler, B. H. (1992) Reduction of potassium conductances mediated by metabotropic glutamate receptors in rat CA3 pyramidal cells does not require protein kinase C or protein kinase A. *Eur. J. Neurosci.* **4,** 792–797.

Gilman, A. G. (1987) G proteins, transducers of receptor-generated signals. *Annu. Rev. Biochem.* **56,** 615–649.

Glaum, S. R. and Miller, R. J. (1992) Metabotropic glutamate receptors mediate excitatory transmission in the nucleus of the solitary tract. *J. Neurosci.* **12,** 2251–2258.

Glaum, S. R., Slater, N. T., Rossi, D. J., and Miller, R. J. (1992) Role of metabotropic glutamate ACPD) receptors at the parallel fiber-Purkinje cell synapse. *J. Neurophysiol.* **68,** 1453–1461.

Greene, C., Schwindt, P., and Crill, W. (1992) Metabotropic receptor mediated after depolarization in neocortical neurons. *Eur. J. Pharmacol.* **226,** 279–280.

Guérineau, N. C., Gähwiler, B. H., and Gerber, U. (1994) G-proteins mediate reduction of resting K⁺ current by metabotropic glutamate and muscarinic receptors in rat CA3 cells. *J. Physiol.* **474,** 27–33.

Halliwell, J. V. and Adams, P. R. (1982) Voltage-clamp analysis of muscarinic excitation in hippocampal neurons. *Brain Res.* **250,** 71–92.

Harvey, J. and Collingridge, G. L. (1993) Signal transduction pathways involved in the acute potentiation of NMDA responses by 1*S*,3*R*-ACPD in rat hippocampal slices. *Br. J. Pharmacol.* **109,** 1085–1090.

Hu, G.-Y. and Storm, J. F. (1992) Excitatory amino acids acting on metabotropic glutamate receptors broaden the action potential in hippocampal neurons. *Brain Res.* **568,** 339–344.

Inoue, T., Miyakawa, H., Ito, K., Mikoshiba, K., and Kato, H. (1992) A hyperpolarizing response induced by glutamate in mouse cerebellar Purkinje cells. *Neurosci. Res.* **15,** 265–271.

Irving, A. J., Schofield, J. G., Watkins, J. C., Sunter, D. C., and Collingridge, G. L. (1990) 1*S*,3*R*-ACPD stimulates and L-AP3 blocks Ca²⁺ mobilization in rat cerebellar neurons. *Eur. J. Pharmacol.* **186,** 363–365.

Ito, M. and Karachot, L. (1990) Messengers mediating long-term desensitization in cerebellar Purkinje cells. *NeuroReport* **1,** 129–132.

Jan, L. Y. and Jan, Y. N. (1992) Tracing the roots of ion channels. *Cell* **69,** 715–718.

Kaczmarek, L. K. and Levitan, I. B. (1987) *Neuromodulation: The Biochemical Control of Neuronal Excitability* (Oxford University Press, Oxford, UK).

Katada, T. and Ui, M. (1982) Direct modification of the membrane adenylyl cyclase system by islet-activating protein due to ADP-ribosylation of a membrane protein. *Proc. Natl. Acad. Sci. USA* **79,** 3129–3133.

Kinney, G. A. and Slater, N. T. (1992) Potentiation of mossy fiber-evoked EPSPs in turtle cerebellar Purkinje cells by the metabotropic glutamate receptor agonist 1*S*,3*R*-ACPD. *J. Neurophysiol.* **67,** 1006–1008.

Kubo, Y., Reuveny, E., Slesinger, P. A., Jan, Y. N., and Jan, L. Y. (1993) Primary structure and functional expression of a rat G-protein-coupled muscarinic potassium channel. *Nature* **364,** 802–806.

Lancaster, B., and Adams, P. R. (1986) Calcium-dependent current generating the after hyperpolarization of hippocampal neurons. *J. Neurophysiol.* **55,** 1268–1282.

Lancaster, B., Nicoll, R. A., and Perkel, D. J. (1991) Calcium activates two types of potassium channels in rat hippocampal neurons in culture. *J. Neurosci.* **11,** 23–30.

Lester, R. A. J. and Jahr, C. E. (1990) Quisqualate receptor-mediated depression of calcium currents in hippocampal neurons. *Neuron* **4,** 741–749.

Lüthi, A., Gähwiler, B. H., and Gerber, U. (1993) Interaction between ionotropic and metabotropic glutamate receptors in the hippocampus. *Experientia* **49,** A74.

Madison, D. V. and Nicoll, R. A. (1986) Cyclic adenosine 3',5'-monophosphate mediates -receptor actions of noradrenaline in rat hippocampal pyramidal cells. *J. Physiol. (Lond.)* **372,** 245–259.

Malenka, R. C., Madison, D. V., Andrade, R., and Nicoll, R. A. (1986) Phorbol esters mimic some cholinergic actions in hippocampal pyramidal neurons. *J. Neurosci.* **6,** 475–480.

Manzoni, O., Fagni, L., Pin, J.-P., Rassendren, F., Poulat, F., Sladeczek, F., and Bockaert, J. (1990) *(trans)*-1-amino-cyclopentyl-1,3-dicarboxylate stimulates quisqualate phosphoinositide-coupled receptors but not ionotropic glutamate receptors in striatal neurons and *Xenopus* oocytes. *Mol. Pharmacol.* **38,** 1–6.

Masu, M., Tanable, Y., Tsuchida, K., Shigemoto, R., and Nakanishi, S. (1991) Sequence and expression of a metabotropic glutamate receptor. *Nature* **349,** 760–765.

McCormick, D. A. (1991) Cellular mechanisms underlying cholinergic and noradrenergic modulation of neuronal firing mode in the cat and guinea pig dorsal lateral geniculate nucleus. *J. Neurosci.* **12,** 278–289.

McCormick, D. A. and von Krosigk, M. (1992) Corticothalamic activation modulates thalamic firing through glutamate "metabotropic" receptors. *Proc. Natl. Acad. Sci. USA* **89,** 2774–2778.

Miles, R. and Poncer, J.-C. (1993) Metabotropic glutamate receptors mediate a post-tetanic excitation of guinea-pig hippocampal inhibitory neurones. *J. Physiol. Lond.* **463,** 461–473.

Miller, R. J. (1991) Metabotropic excitatory amino acid receptors reveal their true colors. *TiPS* **146,** 365–367.

Müller, W., Petrozzino, J. J., Griffith, L. C. Dahno, W., and Connor, J. A. (1992) Specific involvement of Ca^{2+}-calmodulin kinase II in cholinergic modulation of neuronal responsiveness. *J. Neurophysiol.* **68,** 2264–2269.

Nakanishi, S. (1992) Molecular diversity of glutamate receptors and implications for brain function. *Science* **258**, 597–603.

Palmer, E., Monaghan, D. T., and Cotman, C. W. (1989) Trans-ACPD, a selective agonist of the phosphoinositide-coupled excitatory amino acid receptor. *Eur. J. Pharmacol.* **166**, 585–587.

Rüegg, U. T. and Burgess, G. M. (1989) Staurosporine, K-252 and UCN-01, potent but nonspecific inhibitors of protein kinases. *TiPS* **10**, 218–220.

Sahara, Y. and Westbrook G. L. (1993) Modulation of calcium currents by a metabotropic glutamate receptor involves fast and slow kinetic components in cultured hippocampal neurons. *J. Neurosci.* **13**, 3041–3050.

Sayer, R. J., Schwindt, P. C., and Crill, W. E. (1992) Metabotropic glutamate receptor-mediated suppression of L-type calcium current in acutely isolated neocortical neurons. *J. Neurophysiol* **68**, 833–842.

Schoepp, D., Bockaert, J., and Sladeczek, F. (1990) Pharmacological and functional characteristics of metabotropic excitatory amino acid receptors. *TiPS* **11**, 508–515.

Schoepp, D. D. and Conn, P. J. (1993) Metabotropic glutamate receptors in brain function and pathology. *TiPS* **14**, 13–20.

Schwartz, E. A. (1993) L-glutamate conditionally modulates the K⁺ current of Müller glial cells. *Neuron* **10**, 1141–1149.

Staub, C., Vranesic, I., and Knöpfel, T. (1992) Responses to metabotropic glutamate receptor activation in cerebellar Purkinje cells: induction of an inward current. *Eur. J. Neurosci.* **4**, 832–839.

Sternweis, P. C. and Smrcka, A. V. (1992) Regulation of phospholipase C by G proteins. *TIBS* **17**, 502–506.

Stratton, K. R., Worley, P. F., and Baraban, J. M. (1989) Excitation of hippocampal neurons by stimulation of glutamate Qp receptors. *Eur. J. Pharmacol.* **173**, 235–237.

Stratton, K. R., Worley, P. F., and Baraban, J. M. (1990) Pharmacological characterization of phosphoinositide-linked glutamate receptor excitation of hippocampal neurons. *Eur. J. Pharmacol.* **186**, 357–361.

Sugiyama, H., Ito, I., and Hirono, C. (1987) A new type of glutamate receptor linked to inositol phospholipid metabolism. *Nature* **325**, 531–533.

Sugiyama, H., Ito, I., and Watanabe, M. (1989) Glutamate receptor subtypes may be classified into two major categories, A study on *Xenopus* oocytes injected with rat brain mRNA. *Neuron* **3**, 129–132.

Swartz, K. J. and Bean, B. P. (1992) Inhibition of calcium channels in rat CA3 pyramidal neurons by a metabotropic glutamate receptor. *J. Neurosci.* **12**, 4358–4371.

Takagi, H., Takimizu, H., Yoshioka, T., Suzuki, N., and Kudo, Y. (1992) Delayed appearance of a G-protein coupled signal transduction system in cerebellar Purkinje cell dendrites. *Neurosci. Res.* **15**, 206–212.

Tang, W.-J. and Gilman, A. G. (1991) Type-specific regulation of adenylyl cyclase by G-protein βγ subunits. *Science* **254**, 1500–1503.

Trombley, P. Q. and Westbrook, G. L. (1992) L-AP4 inhibits calcium currents and synaptic transmission via a G-protein-coupled glutamate receptor. *J. Neurosci.* **12**, 2043–2050.

Trussell, L. O. and Jackson, M. B. (1987) Dependence of an adenosine-activated potassium current on a GTP-binding protein in mammalian central neurons. *J. Neurosci.* **7,** 3306–3316.

Vranesic, I., Batchelor, A., Gähwiler, B. H., Garthwaite, J., Staub, C., and Knöpfel, T. (1991) Trans-ACPD-induced Ca^{2+} signals in cerebellar Purkinje cells. *NeuroReport* **2,** 759–762.

Vranesic, I., Staub, C., and Knöpfel, T. (1993) Activation of metabotropic glutamate receptors induces an outward current which is potentiated by methylxanthines in rat cerebellar Purkinje cells. *Neurosci. Res.* **16,** 209–215.

Wang, Z. and McCormick, D. A. (1993) Control of firing mode of corticotectal and corticopontine layer V burst-generating neurons by norepinephrine, acetylcholine, and 1S,3R-ACPD. *J. Neurosci.* **13,** 2199–2216.

Yatani, A., Codina, J., Imoto, Y., Reeves, J. P., Birnbaumer, L., and Brown, A. M. (1987) A G-protein directly regulates mammalian cardiac calcium channels. *Science* **238,** 1288–1292.

Yool, A. J., Krieger, R. M., and Gruol, D. L. (1992) Multiple ionic mechanisms are activated by the potent agonist quisqualate in cultured cerebellar Purkinje neurons. *Brain Res.* **573,** 83–94.

Zegarra-Moran, O., and Moran, O. (1993) Modulation of voltage-dependent calcium channels by glutamate in rat cerebellar granule cells. *Exp. Brain Res.* **95,** 65–69.

Zheng, F. and Gallagher, J. P. (1991) *Trans*-ACPD (*trans*-D,L-1-amino-1,3-cyclopentanedicarboxylic acid) elicited oscillation of membrane potentials in rat dorsolateral septal nucleus neurons recorded intracellularly in vitro. *Neurosci. Lett.* **125,** 147–150.

Zheng, F. and Gallagher, J. P. (1992a) Metabotropic glutamate receptor agonists potentiate a slow after hyperpolarization in CNS neurons. *NeuroReport* **3,** 622–624.

Zheng, F. and Gallagher, J. P. (1992b) Burst firing of rat septal neurons by 1S,3R-ACPD requires influx of extracellular calcium. *Eur. J. Pharmacol.* **211,** 281–282.

Zorumski, C. F. and Thio, L. L. (1992) Properties of vertebrate glutamate receptors: calcium mobilization and desensitization. *Prog. Neurobiol.* **39,** 295–336.

CHAPTER 6

Acute Regulation of Synaptic Transmission by Metabotropic Glutamate Receptors

Steven R. Glaum and Richard J. Miller

1. Introduction

Two principal classes of glutamate receptors have been identified: (1) ligand-gated ion channels and (2) G-protein-coupled
"metabotropic" receptors (Sugiyama et al., 1989). Activation of
ionotropic α-amino-3-hydroxy-5-methyl-4-isoxazole-propionic acid
(AMPA), kainate (KA), and N-methyl-D-aspartate (NMDA) receptors
represents the principal route of fast excitatory transmission in the
CNS. However, it is becoming increasingly clear that synaptic transmission also appears to be influenced by the actions of glutamate on
metabotropic glutamate receptors (mGluRs) at both pre- and postsynaptic sites. At least seven mGluR subtypes (mGluR1–7) plus several
splice varients have been identified by molecular biological methods
(*see* Chapter 1). Expression of these receptors in a variety of cell
types has shown that they are capable of interacting with most of
the commonly recognized second-messenger systems. As detailed

The Metabotropic Glutamate Receptors Eds.: P. J. Conn and J. Patel
© 1994 Humana Press Inc., Totowa, NJ

elsewhere in this volume, each expressed mGluR subtype also displays unique pharmacological specificity and shows a particular preference for one of the effector systems (Nakajima et al., 1993; Tanabe et al., 1993). However, which mGluRs mediate the various acute effects of mGluR activation on synaptic transmission and the underlying mechanisms are still poorly understood. In this chapter, we will review the current understanding regarding acute regulation of synaptic transmission by mGluRs and examine possible mechanisms of this regulation.

2. Presynaptic Interactions

2.1. Glutamate Autoreceptors

Much of our current understanding of glutamate receptor physiology has come from the use of pharmacological tools that selectively mimic or block one of glutamate's many effects in the CNS. Of particular importance to the present discussion are the observations of Koerner and Cotman (1981) that the L isomer of 2-amino-4-phosphonobutyrate (L-AP4) can act as a potent synaptic depressant in the dentate outer molecular layer of the guinea-pig hippocampus. L-AP4 was similarly active at inhibiting excitatory transmission in guinea-pig hippocampal CA1 neurons, but was less clearly effective in the CA1 of the adult rat (Lanthorn et al., 1984). Davies and Watkins (1982) extended these observations to other regions of the CNS by demonstrating depression of transmission by L-AP4 in the cat spinal cord. In the above instances, the excitatory transmission being suppressed was the result of the actions of glutamate on postsynaptic ionotropic glutamate receptors. It was initially thought that L-AP4 was acting as a postsynaptic glutamate antagonist. However, Harris and Cotman (1983) altered this view following the publication of results that suggested a presynaptic site of action in the lateral perforant pathway-dentate gyrus, thereby suggesting that L-AP4 was actually acting as an agonist on an inhibitory glutamatergic "autoreceptor." It is likely that these effects of L-AP4 are mediated by one of the mGluRs that are selectively sensitive to this agonist (e.g., mGluR 4, 6, or 7; *see* Chapter 1).

Our understanding of mGluR-mediated events in the CNS has been substantially advanced by the introduction of the rigid structural glutamate analog 1-aminocyclopentane-1,3-dicarboxylic acid (ACPD), and in particular its active (1S,3R) isomer ([1S,3R]-ACPD), which acts as a selective mGluR agonist. For example, Palmer et al. (1989) and Desai and Conn (1991) were able to demonstrate that ACPD activation of mGluR in the hippocampus affected both phosphatidylinositol (PI) turnover and synaptic transmission in vitro. Both ACPD and L-AP4 have also been shown to inhibit EPSCs recorded in CA1 presynaptically following stimulation of Schaffer collateral/ commissural afferents in neonatal rats (Baskys and Malenka, 1991). These data indicate that glutamatergic synapses in the hippocampus contain one or more presynaptic inhibitory mGluR autoreceptors that are activated by L-AP4 and ACPD.

More recently, widespread inhibitory effects of (1S,3R)-ACPD on glutamatergic transmission have been described. As shown in Fig. 1, ACPD is able to inhibit monosynaptic glutamatergic EPSP/EPSCs reversibly in rat striatum (Lovinger, 1991), and also at the parallel fiber-Purkinje cell synapse (Crepel et al., 1991; Glaum et al., 1992), in rat motorneurons (Pook et al., 1992), in the basolateral amygdala (Rainnie and Shinnick-Gallagher, 1992), and in the nucleus of the solitary tract (NTS) (Glaum and Miller, 1992), among other sites. In the latter case, using a selective mGluR antagonist, we have recently demonstrated that endogenously released glutamate does appear to inhibit its own release via activation of a presynaptic mGluR autoreceptor (Glaum and Miller, 1993b). This broad distribution of inhibitory effects of ACPD on glutamate release may indicate that inhibitory mGluR autoreceptors are a prominent feature of the glutamatergic synapse in the CNS.

As mentioned above, there is now pharmacological evidence that the "L-AP4" autoreceptor in many systems may be mGluR4, although mGluR7 may play a similar role in discrete brain areas (Westbrook et al., 1993). mGluR4 has been shown to inhibit forskolin-stimulated adenylyl cyclase in vitro (Thomsen et al., 1992). However, it has not yet been demonstrated that the ability of either L-AP4 or (1S,3R)-ACPD to inhibit glutamatergic transmission is

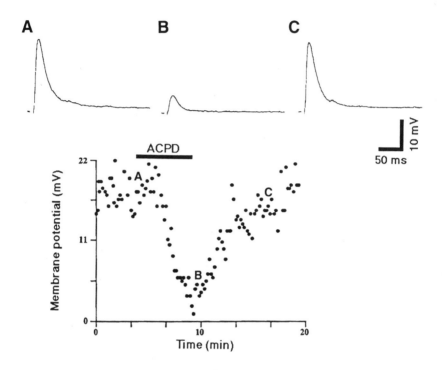

Fig. 1. Synaptic depression mediated by ACPD is observed in rat neostriatal neurons recorded in slices with whole-cell patch electrodes. Bath applied *trans*-ACPD (100 μ*M*) transiently depresses monosynaptically evoked EPSPs. Waveforms above graph show averages of five EPSPs recorded at the indicated times during the experiment. Data were obtained from a 3-wk-old rat. (Figure courtesy of D. Lovinger.)

actually mediated by an effect on adenylyl cyclase. Autoreceptor effects of (1S,3R)-ACPD also appear to be independent of mGluR-mediated increases in PI turnover and Ca²⁺ mobilization, the initial mechanism assigned to mGluR activation (Sladeczek et al., 1985; Nicoletti et al., 1986; Glaum et al., 1991). This is also consistent with the relatively low affinity of L-AP4 and ACPD for cloned mGluRs that are preferentially linked to phospholipase C (Aramori and Nakanishi, 1992; Abe et al., 1992). Moreover, the ability of ACPD to inhibit EPSPs acutely at the cerebellar parallel fiber-Purkinje cell synapse (Crepel et al., 1991; Glaum et al., 1992) or the hippocampal Schaffer collateral–CA1 synapse (Boss et al., 1992) is insensitive to the effects of L-2-amino-3-phosphonopropionate (L-AP3), a compound that blocks the ability of ACPD to enhance PI turnover in many systems (Palmer et al., 1989; Schoepp, 1993; however, *see* Nadler et al., 1993).

What second-messenger systems might therefore be activated by mGluR autoreceptors? A likely candidate for the mechanism by which presynaptic mGluRs inhibit transmitter release comes from the recent observation that their activation can inhibit voltage-gated Ca^{2+} channels in a pertussis toxin-sensitive manner (Lester and Jahr, 1990; Swartz and Bean, 1992; Sahara and Westbrook, 1993; Zeilhofer et al., 1993), a property shared by a number of other G-protein-linked receptors (Milligan, 1993). Activation of mGluRs has been shown to inhibit N-type channels (Swartz and Bean, 1992; Sahara and Westbrook, 1993) and P/Q-type channels (Randall et al., 1993), both of which have been implicated in neurotransmitter release in the CNS. Recent results suggest that this Ca^{2+} channel inhibition may result from a negative interaction between mGluRs and PKC or a PKC substrate (Swartz, 1993; Swartz et al., 1993). Alternatively, mGluR regulation of Ca^{2+} and other channels (*see* Section 2.3.) may occur via a cGMP-dependent mechanism (Glaum and Miller, 1993c) or by more direct coupling to channels via a G-protein. Finally, presynaptic mGluRs might indirectly reduce Ca^{2+} currents and neurotransmitter release by activation of presynaptic K^+ channels (Sladeczek et al., 1993).

Before examining presynaptic mGluR interactions with non-glutamatergic neurotransmitter systems, it should also be noted that the mGluR autoreceptor agonist L-AP4 has been shown to also activate a postsynaptic receptor subtype, the recently cloned mGluR6, that mediates neurotransmission between retinal photoreceptor cells and ON-bipolar neurons (Nakajima et al., 1993). Noteably, this receptor appears to regulate a cGMP-gated conductance (Nawy and Jahr, 1991; *see* Chapter 8), further suggesting that mGluRs may couple to second messenger systems other than PI turnover and adenylyl cyclase.

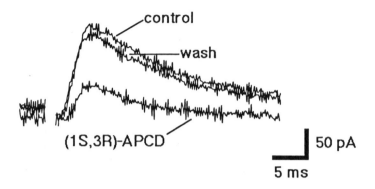

Fig. 2. Monosynaptically evoked IPSCs are reversibly inhibited in the rat NTS by (1S,3R)-ACPD (50 μ*M*). IPSCs were evoked by electrical stimulation in the region of the tractus solitarius at 0.1 Hz in the presence of the ionotropic glutamate receptor antagonists 6,7-dinitroquinoxaline-2,3-dione (10 μ*M*) and D-amino-5-phosphonopentanoic acid (50 μ*M*), and recorded in an NTS neuron medial to the tractus solitarius and adjacent to the area postrema. Whole-cell voltage-clamp recordings (V_{hold} = –50) were made with K-gluconate-containing electrodes (Glaum and Miller, 1992).

2.2. Inhibitory Heteroreceptors

In addition to the widespread inhibitory actions of mGluR activation on glutamatergic transmission, a more limited number of studies have demonstrated presynaptic inhibitory effects of mGluRs on other neurotransmitter systems, particularly GABA. For example, mGluR activation inhibits GABA-mediated inhibitory synaptic transmission in the hippocampus (Desai and Conn, 1991; Pacelli and Kelso, 1991). As illustrated in Fig. 2, (1S,3R)-ACPD also produces an acute inhibition of pharmacologically isolated, monosynaptic $GABA_A$-mediated IPSCs in the NTS (Glaum and Miller, 1992).

It should be noted, however, that the actions of (1S,3R)-ACPD on GABAergic transmission in the NTS may be partly mediated by postsynaptic mGluRs (*see* Section 3.). In contrast, Hayashi et al. (1993) have demonstrated that presynaptic mGluRs, possibly mGluR2, can inhibit the release of GABA from granule cells in the rat accessory olfactory bulb. It is likely that additional examples of presynaptic mGluR-mediated inhibition of nonglutamatergic transmission will be described in the near future.

2.3. Facilitatory Heteroreceptors

Posttetanic potentiation (PTP) is a widely observed phenomenon at many synapses. An acute increase in presynaptic intracellular free Ca^{2+} ($[Ca^{2+}]_i$) appears to be an important component of PTP (Mulkey and Zucker, 1992). Could mGluRs, some of which mobilize $[Ca^{2+}]_i$ (e.g., mGluRs 1 and 5), also facilitate transmission? Glutamate has been demonstrated to enhance the release of a wide variety of neurotransmitters in vitro (reviewed in Ruzicka and Jhamandas, 1993), yet it remains to be demonstrated which of these effects might be mediated via an interaction with an mGluR. Clearly, (1S,3R)-ACPD can produce direct excitatory effects on a number of neurons receiving glutamatergic inputs, as described in the following sections, which presumably would enhance ongoing transmission. However, recent reports also suggest that glutamate can enhance its own release in the rat cortex via an interaction with a presynaptic mGluR (Coffey et al., 1993; Sanchez-Prieto et al., 1993). This action appears to be related to local release of arachidonic acid and is mediated by activation of PKC (Coffey et al., 1993). The effect is inhibited by L-AP3 (Sanchez-Prieto et al., 1993), which inhibits mGluR in some systems (Irving et al., 1990; Crepel et al., 1991). Similarly, L-AP4 and (1S,3R)-ACPD reduce paired-pulse depression in the hippocampus (i.e., a net facilitatory effect), suggestive of actions beyond the previously mentioned inhibitory activity of these compounds on glutamate release (Kahle and Cotman, 1993).

A clear demonstration of a role for mGluRs in the facilitation of transmission at nonglutamatergic synapses has proven to be more elusive. Stelzer and Wong (1989) have demonstrated that GABAergic transmission in the rat hippocampus is enhanced by mGluR activation, but the locus of this effect is unclear. Recently, Miles and Poncer (1993) have provided evidence that mGluRs on guinea-pig hippocampal inhibitory neurons are activated during tetanic stimulation and mediate a posttetanic increase in excitability, thereby facilitating inhibitory transmission. Thus, it appears that mGluRs can facilitate nonglutamatergic transmission in vitro. Although a definitive demonstration of this phenomenon in vivo has yet to be reported, Sacaan et al. (1992) have suggested that the behavioral effects of (1S,3R)-

ACPD injected into striatum may be mediated by facilitation of dopaminergic transmission. Although intriguing, it remains unclear if this effect takes place via a facilitory heteroreceptor at the terminal of the dopaminergic neuron, by a decrease in the activity of an inhibitory pathway, or via a more generalized increase in striatal neuronal excitability. Notably, Schoepp (1993) has found that (1S,3R)-ACPD failed to modulate directly the release of dopamine in vitro, a result favoring the latter hypothesis.

The mechanisms underlying facilitation of neurotransmitter release by mGluRs are still largely speculative. An effect on Ca^{2+} channels may be involved. For example, (1S,3R)-ACPD has been shown to potentiate an L-type Ca^{2+} current in cerebellar granule cells (Bossu et al., 1993). However as indicated above, (1S,3R)-ACPD also appears to suppress the Ca^{2+} current through N- and P/Q-type Ca^{2+} channels in several instances (Swartz and Bean, 1992; Randall et al., 1993), and these Ca^{2+} channels are most closely associated with neurotransmitter release (Miller, 1990). The use of selective dihydropyridine and peptide antagonists of Ca^{2+} channels may provide further insight into these mechanisms. One should also consider that these effects may not only be the result of direct actions of mGluRs at or near presynaptic terminals, but may also represent mGluR-mediated interactions with other widespread neuromodulatory substances, such as adenosine (Caciagli et al., 1993; Winder and Conn, 1993; Zhu and Krnjevic, 1993) or arachidonic acid (Sanchez-Prieto et al., 1993). Furthermore, as described in the following sections, postsynaptic mGluRs may also be capable of modulating the activity of receptors for glutamate and other transmitter systems.

3. Postsynaptic Interactions

mGluRs can also clearly regulate synaptic transmission at postsynaptic sites through modulatory effects on voltage- and ligand-gated ion channels. In particular, effects on K^+ conductances, ionotropic glutamate, and GABA receptors have been observed. The following sections will focus on the contribution postsynaptic mGluRs make to acute regulation of synaptic transmission and begin to explore the underlying processes responsible for these effects.

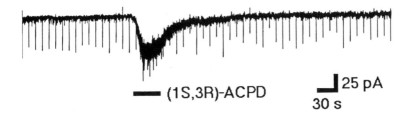

Fig. 3. (1S,3R)-ACPD (50 μM) produces a transient inward current and reduces monosynaptically evoked EPSCs in the NTS. EPSCs were evoked from the region of the tractus solitarius in the presence of bicuculline (10 μM) and D-amino-5-phosphonopentanoic acid (50 μM), and recorded with K-gluconate-filled whole-cell electrodes (V_{hold} = –55 mV).

3.1. mGluR-Mediated Currents

A prominent observation in the CNS following application of (1S,3R)-ACPD is an effect on one or more postsynaptic K^+ currents. A reduction in a Ba^{2+}-sensitive K^+ conductance is observed in NTS neurons, as illustrated in Fig. 3, (Glaum and Miller, 1992, 1993b; Priddy et al., 1992), hippocampal cells (Charpak et al., 1990; Desai and Conn, 1991; Gerber et al., 1993), and in the thalamus (McCormick and von Krosigk, 1992).

In the NTS, high-frequency electrical stimulation of glutamatergic afferents in the presence of ionotropic glutamate antagonists can activate the postsynaptic mGluR(s) acting on this K^+ current (Glaum and Miller, 1992). Similar results have recently been obtained in the hippocampus (Gerber et al., 1993) and cerebellum (Batchelor and Garthwaite, 1993). By reducing this tonic conductance, neuronal excitability can be enhanced, in some cases bringing the resting membrane potential into the range of spontaneous action potential firing. Similarly, activation of the corticothalamic pathway has been shown to activate postsynaptic mGluRs, resulting in a brief increase in the excitability of thalamic neurons to subsequent stimulation (McCormick and von Krosigk, 1992). These data support the notion that mGluRs play an important physiological role in mediating excitatory transmission in the CNS, particularly during periods of increased afferent activity.

The nature of the K^+ conductances modulated by postsynaptic mGluRs seems to vary depending on the particular cell type and species being examined. For example, enhancement of hippocampal, but not NTS excitability by mGluRs is attributable to suppression of I_M and the Ca^{2+}-activated K^+ current underlying I_{AHP} (Stratton et al., 1989; Charpak et al., 1990; Hu and Storm, 1991; Glaum and Miller, 1992; Gerber et al., 1993). A sustained membrane depolarization attributable to a novel K^+-mediated inward tail current (I_{ADP}) is observed in guinea-pig olfactory cortical neurons (Constanti and Libri, 1992). A similar increase in I_{ADP} is observed in rat neocortical neurons (Greene et al., 1992). mGluR-mediated burst firing in rat dorsolateral septal nucleus neurons has been demonstrated to be blocked by inorganic calcium-channel blockers (Zheng and Gallagher, 1992), suggesting a requirement for Ca^{2+} influx. In Aplysia neurons, (1S,3R)-ACPD activates two currents, a rapidly desensitizing Cl^- current and a nondesensitizing outwardly rectifying K^+ current (Katz and Levitan, 1993).

Special note should be given to the postsynaptic effects of mGluR activation in the cerebellum, where multiple postsynaptic effects of (1S,3R)-ACPD have been described. For example, (1S,3R)-ACPD produces a transient depolarization leading to burst firing and Ca^{2+} spikes in Purkinje cells (Crepel et al., 1991; Glaum et al., 1992; Yool et al., 1992). These acute changes in membrane excitability may underlie more sustained alterations in synaptic efficacy, such as those relating to long-term depression at the parallel fiber-Purkinje cell synapse, as detailed elsewhere in this volume. (1S,3R)-ACPD also produces both a transient inward (Vranesic et al., 1991) and slow outward currents (Vranesic et al., 1993) in Purkinje cells. In cerebellar granule cells, mGluR-mediated PI hydrolysis can liberate Ca^{2+} from intracellular stores and activate $I_{K^+(Ca2+)}$ (Aronica et al., 1993). Similar effects may also be occurring in Purkinje neurons (Vranesic et al., 1991; Staub et al., 1992). However, intracellular Ca^{2+} mobilization by an mGluR-mediated mechanism in Purkinje neurons has not been consistently observed (Llano et al., 1991; Glaum et al., 1992). In addition, identification of the ionic conductances underlying these mGluR-mediated currents has been problematic. These difficulties in identifying mGluR-mediated mechanisms in Purkinje neurons may be the result in part of the problem of adequately voltage-clamping these cells, which have extensive dendritic processes. For example, Vranesic et al. (1993) have characterized a slow outward current in Purkinje cells that is inhibited by (1S,3R)-ACPD and is associated with a decrease in membrane conductance, but not with a detectable change in $[Ca^{2+}]_i$. This current is therefore presumably unrelated to $[Ca^{2+}]_i$ mobilization. Notably, inhibition of this current by (1S,3R)-ACPD is enhanced in the presence of methylxanthines (Vranesic et al., 1993), indicating a potential role for cAMP. Additional effects of (1S,3R)-ACPD in Purkinje cells include an mGluR-mediated inhibition of a K^+ "leak" current and a transient inward conductance that may be mediated by reversal of a transmembrane Na^+/Ca^{2+} exchanger following mobilization of $[Ca^{2+}]_i$ (Konnerth et al., 1990; Crepel et al., 1991; Linden and Connor, 1991; Vranesic et al., 1991, 1993; Staub et al., 1992). The depolarizing influence of (1S,3R)-ACPD on Purkinje neurons makes segregation of Ca^{2+} changes associated with mobilization and those associated with influx through voltage-gated Ca^{2+} channels difficult. The presence of multiple (1S,3R)-ACPD responses in Purkinje neurons also makes the isolation of individual ion conductances problematic. Selective activation or inhibition of mGluR-coupled second-messenger systems may make it possible to examine these currents in isolation. Alternatively, the use of the more selective mGluR agonists and antagonists described at the end of this chapter may assist in describing underlying mechanisms.

The preceding array of modulatory effects produced by mGluR is likely to grow as additional regions of the brain are investigated. It is unclear at present which mGluR(s) are responsible for these effects. For example, mobilization of intracellular Ca^{2+} has been demonstrated in many systems, particularly the hippocampus (Palmer et al., 1989; Glaum et al., 1991), striatum (Ambrosini and Meldolesi, 1989; Manzoni et al., 1990), and cerebellum (Irving et al., 1990; Manzoni et al., 1992). These responses are likely to be mediated by activation of mGluRs 1 and 5, which have been shown to couple to phospholipase C when expressed in vitro (Schoepp, 1993;

see Chapter 1). However, although some of the membrane effects of (1S,3R)-ACPD could be secondary to effects on $[Ca^{2+}]_i$, it is notable that the electrophysiological effects of (1S,3R)-ACPD (and presumably the underlying mGluRs) appear to undergo relatively few developmental changes. In contrast, mGluR-mediated effects on PI turnover and the underlying mGluRs are highly plastic, as recently described in the hippocampus (Boss et al., 1992), cerebellar granule cells (Aronica et al., 1993), Purkinje cells (Catania et al., 1993), and elsewhere in the CNS (Martin et al., 1993). With the exception of cerebellar granule cells, changes in $[Ca^{2+}]_i$ monitored with the fluorescent indicator fura-2 have also proven difficult to correlate with mGluR-mediated electrophysiological effects. Thus, mGluRs 1 and 5 may generally subserve other functions than those described in the preceding section or may activate additional effector systems to those already demonstrated.

3.2. Ionotropic Glutamate Receptors

One of the more interesting observations regarding acute regulation of synaptic transmission by mGluRs is that, in contrast to autoreceptor inhibition of glutamate release, postsynaptic ionotropic glutamate receptor activity is often enhanced by (1S,3R)-ACPD. Experiments performed in our own laboratory on brain slices of the rat spinal cord have shown that currents mediated by exogenous AMPA, KA, and NMDA are enhanced in the presence of (1S,3R)-ACPD (Bleakman et al., 1992). As illustrated in Fig. 4, AMPA currents are also transiently enhanced in NTS neurons (Glaum and Miller, 1993a) and rat cerebellar Purkinje cells (Glaum et al., 1992). On the other hand, only NMDA responses are potentiated in turtle Purkinje neurons and in rat granule cells, as illustrated in Fig. 5 (Aronica et al., 1993; Kinney and Slater, 1993).

A similar potentiation of NMDA, but not AMPA currents has been observed in hippocampal CA1 neurons (Aniksztejn et al., 1991, Ben-Ari et al., 1993). An increase in NMDA-induced brain injury in the presence of mGluR agonists has also been reported (McDonald and Schoepp, 1993). As described below, this may have important implications for both acute and long-term modulation of synaptic efficacy.

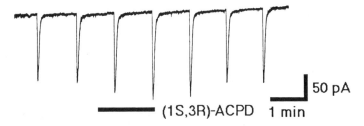

Fig. 4. Inward currents evoked by pressure ejection of AMPA onto an NTS neuron are reversibly potentiated in the presence of 10 μ*M* (1S,3R)-ACPD. AMPA currents were evoked in the presence of bicuculline (10 μ*M*) and D-amino-5-phosphonopentanoic acid (50 μ*M*). Whole-cell recordings were made with K-gluconate electrodes (V$_{hold}$ = –50 mV).

The mechanism by which mGluRs potentiate postsynaptic ionotropic glutamate receptor responses is poorly understood. In addition to effects on PI turnover, a number of additional actions on second-messenger systems have been described (*see* Chapter 3), including modulation of adenylyl cyclase (Cartmell et al., 1992; Winder and Conn, 1993), guanylyl cyclase (Glaum and Miller, 1993c; Irving et al., 1993; Mori-Okamoto et al., 1993), and the liberation of putative gaseous second messengers such as nitric oxide (Ito and Karachot, 1990; Okada 1992) or carbon monoxide (CO) (Marks et al., 1991; Barinaga, 1993; Maines, 1993; Stevens and Wang, 1993; Verma et al., 1993). mGluRs may also activate phospholipase D, although the role this mechanism might play in acute synaptic regulation is unclear (Boss and Conn, 1993; Holler et al., 1993). We have recently demonstrated the likely involvement of mGluR-mediated release of CO in acute regulation of AMPA receptor activity in the NTS (Glaum and Miller, 1993c). The intracellular release of CO may then activate a soluble guanylyl cyclase (Glaum and Miller, 1993c). Although the mechanism by which AMPA receptors in the NTS are upregulated is unclear at present, it may involve a transient dephosphorylation event, since the effects of (1S,3R)-ACPD or the cell-permeant cGMP analog 8Br-cGMP are inhibited in the presence of low concentrations of the phosphatase inhibitor okadaic acid (Glaum and Miller, 1994).

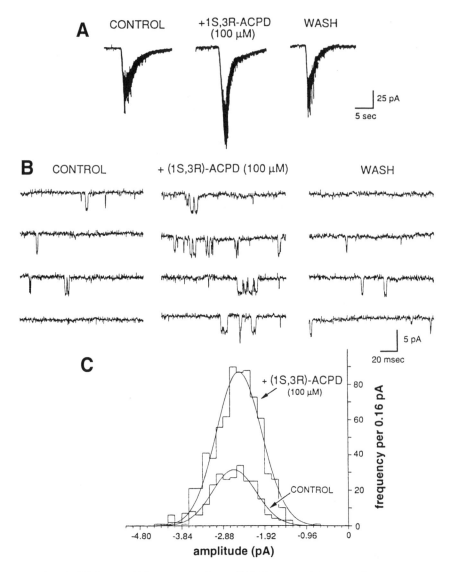

Fig. 5. (A) Reversible potentiation of NMDA sensitivity and spontaneous NMDA receptor-mediated single-channel openings in patch-clamped granule cells in cerebellar slices. Granule cells were recorded using the perforated-patch method in the internal granule cell layer of slices of developing rat cerebellum. A. Bath application of (1S,3R)-ACPD reversibly potentiates the macroscopic current evoked by micropressure ejection of NMDA onto the cell. (B) Representative whole-cell recordings of NMDA receptor-mediated single-channel events in developing granule cells. The opening probability, but not the unitary conductance is potentiated in the presence of (1S,3R)-

As described above, mGluR-mediated potentiation of NMDA currents in the hippocampus has been associated with an increase in both the mean open time and open state probability of the NMDA receptor channel (Kovalchuk et al., 1993). In *Xenopus* oocytes expressing both metabotropic and ionotropic glutamate receptors (Kelso et al., 1992) and in the hippocampus (Ben-Ari et al., 1993), selective potentiation of NMDA currents by quisqualate or ACPD appears to be mediated by a PKC-mediated mechanism (Aniksztejn et al., 1991). This may be of particular importance not only to acute changes in synaptic efficacy, but also to long-term potentiation, as suggested by the recent finding that blocking hippocampal mGluRs prevents LTP in vivo (Riedel and Reymann, 1993; *see* Chapter 7). In terms of mechanism, cloned glutamate receptor subunits appear to contain multiple consensus sites for phosphorylation (Raymond et al., 1993). Thus, mGluR-mediated changes in receptor phosphorylation via kinase and/or phosphatase activation may be a common factor in their modulation.

3.3. Other Neurotransmitter Systems

In addition to mGluR effects on postsynaptic ionotropic glutamate receptors, mGluRs have also been reported to modulate postsynaptic GABA responses in a number of systems. As illustrated in Fig. 6, postsynaptic $GABA_A$ receptor activity in the NTS is reversibly inhibited by (1S,3R)-ACPD (Glaum and Miller, 1993a). As mentioned earlier, (1S,3R)-ACPD inhibits monosynaptic IPSP/IPSCs in the NTS, but the relative contribution of pre- and postsynaptic mGluRs to this observation is unclear at present.

Inhibition of $GABA_A$ responses has also been reported in rabbit hippocampal CA1 neurons (Liu et al., 1993), although this effect is of a longer duration than in the brainstem. On the other hand, as

---

(Fig. 5 *continued*) ACPD. (**C**) Amplitude–frequency histogram for single-channel events in a different cell illustrates the enhanced frequency of channel occurrence, but lack of effect on unitary conductance observed in the presence of (1S,3R)-ACPD. The data were fit with a single Gaussian function. (Figure courtesy of N. T. Slater.)

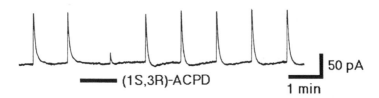

Fig. 6. Outward currents evoked by pressure ejection of muscimol onto NTS neuron are reversibly reduced in the presence of 25 μ*M* (1S,3R)-ACPD. Muscimol currents were evoked in the presence of 6,7-dinitroquinoxaline-2,3-dione (10 μ*M*) and D-amino-5-phosphonopentanoic acid (50 μ*M*). Whole-cell recordings were made with K-gluconate electrodes (V_{hold} = –50 mV).

illustrated in Fig. 7, both GABA- and glycine-mediated inhibition of AMPA-evoked firing in the rat spinal cord is potentiated by (1S,3R)-ACPD in vivo.

The mechanisms responsible for mGluR potentiation of non-glutamatergic transmission are largely unknown at present. In the rat NTS, GABA$_A$ responses appear to undergo complex regulation by mGluRs, apparently involving cGMP activation of one or more phosphatases (Glaum and Miller,1993c,1994). For example, the cell permeant analog 8Br-cGMP mimics the ability of (1S,3R)-ACPD to reduce muscimol currents reversibly (Glaum and Miller, 1993b). The effects of both (1S,3R)-ACPD and 8Br-cGMP are inhibited by the protein phosphatase inhibitor okadaic acid (Glaum and Miller, 1994). Furthermore, in the presence of the protein phosphatase 2B (calcineurin) antagonist FK506, (1S,3R)-ACPD produces reversible potentiation rather than inhibition of muscimol currents (Glaum and Miller, 1994). These data suggest that mGluRs in NTS may influence more than one protein phosphatase, which may provide for either up- or downregulation of inhibitory postsynaptic currents.

4. Discussion

4.1. (1S,3R)-ACPD and Beyond

It has become increasing clear in recent years that mGluRs are present at a wide variety of CNS synapses. As detailed in this chapter, acute modulatory effects of mGluRs on synaptic transmission have

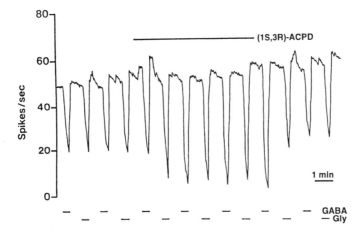

Fig. 7. (1S,3R)-ACPD potentiates the inhibition of firing produced by iontophoretic application of GABA (15 nA) or glycine (Gly, 18 nA) in a spinal neuron recorded in vivo. Baseline firing before and during iontophoretic (1S,3R)-ACPD application (1 nA) was maintained at a fixed level by varying the strength of iontophoretically applied AMPA. All compounds were applied from a seven-barrel microelectrode for the duration indicated by bars on the graph. (Figure courtesy of D. Lodge and A. Bond.)

been widely demonstrated in the CNS. Indeed, mGluRs appear to be present on many, if not all, glutamatergic terminals, where they can act as inhibitory autoreceptors. In addition, we have seen that the selective mGluR agonist (1S,3R)-ACPD can produce acute membrane current responses through its influence on ion channels and produce modulatory effects on postsynaptic ionotropic glutamate receptors and GABA receptors in a variety of neurons. In general, activation of mGluRs appears to be associated with high-frequency synaptic function. This may indicate that postsynaptic mGluRs may not be localized directly within the synaptic cleft, but rather extra-synaptically. The rising number of mGluRs identified by molecular biological methods has raised the question of which mGluRs mediate the aforementioned responses. Unfortunately, (1S,3R)-ACPD distinguishes poorly between the mGluRs. As described in Chapter 2, the development of newer, selective mGluR agonists and antagonists should aid the identification process. For example, selective phenylglycine-derived mGluR antagonists, such as (+)α-methyl-4-carboxyphenylglycine (Birse et al., 1993), have allowed us to discriminate between two mGluR-mediated postsynaptic responses in the NTS (Glaum et al., 1993). More recently, we have used this compound to demonstrate that endogenously released glutamate can inhibit its own release at high-stimulus frequencies via activation of a presynaptic mGluR autoreceptor (Glaum and Miller, 1993b). Newer mGluR agonists, such as (2S,3S,4S)-α-(carboxycyclopropyl)-glycine (L-CCG-I) or 2S,1'R,2'R,3'R)-2-(2,3-dicarboxycyclopropylglycine (DCG-IV), appear to favor activation of specific mGluR subtypes, such as mGluR2 (Nakanishi, 1992; Ishida et al., 1993). More potent and selective antagonists are also on the near horizon. As the mGluR pharmacopia continues to grow, it should be possible to distinguish which mGluRs are involved in a particular synaptic event, determine the underlying mechanisms in vitro, and study more thoroughly the behavioral consequences of mGluR activation in the whole animal.

4.2. Physiological Considerations

Although pharmacological demonstrations of acute mGluR-mediated effects on synaptic transmission are widespread, elucidation of the role of mGluRs in the normal and pathophysiological functions of the CNS has proven elusive. As detailed in this chapter, our investigations in the NTS, for example, have identified responses to mGluR activation that include:

1. A postsynaptic depolarizing response;
2. A presynaptic depression of glutamatergic transmission; and
3. A postsynaptic potentiation of AMPA and depression of $GABA_A$ responses.

All of the effects of (1S,3R)-ACPD appear to be mimicked by endogenously released glutamate (Glaum and Miller, 1993b,c), suggesting that they are of physiological significance. One may speculate that mGluRs in the NTS may help to support excitatory afferent transmission, perhaps vagal in origin, during periods of increased afferent activity. Thus, although presynaptic glutamate release may be reduced via activation of an mGluR autoreceptor, postsynaptic

AMPA receptor sensitivity could be potentiated, GABA receptor sensitivity depressed, and membrane excitability increased via the reduction of a tonic K^+ current. The creation of increasingly selective mGluR agonists and antagonists will benefit our understanding of the physiological roles served by this family of receptors.

Acknowledgments

The authors wish to thank D. Lodge and A. Bond of Eli Lilly, D. Lovinger of Vanderbilt University, N. T. Slater, and D. J. Rossi of Northwestern University for their generous contribution of figures to this chapter.

References

Abe, T., Sugihara, H., Nawa, H., Shigemoto, R., Mizuno, N., and Nakanishi, S. (1992) Molecular characterization of a novel metabotropic glutamate receptor mGluR5 coupled to inositol phosphate/Ca^{2+} signal transduction. *J. Biol. Chem.* **267,** 13,361–13,368.

Ambrosini, A. and Meldolesi., J. (1989) Muscarinic and quisqualate receptor-induced phosphoinositide hydrolysis in primary cultures of striatal and hippocampal neurons. Evidence for differential mechanisms of activation. *J. Neurochem.* **53,** 825–833.

Aniksztejn, L., Bregestovski, P., and Ben-Ari, Y. (1991) Selective activation of quisqualate metabotropic receptor potentiates NMDA but not AMPA responses. *Eur. J. Pharmacol.* **205,** 327,328.

Aramori, I. and Nakanishi, S. (1992) Signal transduction and pharmacological characteristics of a metabotropic glutamate receptor, mGluR1, in transfected CHO cells. *Neuron* **8,** 757–765.

Aronica, E., Condorelli, D. F., Nicoletti, F., Dell'Albani, P., Amico, C., and Balàzs, R. (1993) Metabotropic glutamate receptors in cultured cerebellar granule cells: Developmental profile. *J. Neurochem.* **60,** 559–565.

Barinaga, R. (1993) Carbon monoxide: Killer to brain messenger in one step. *Science* **259,** 309.

Baskys, A. and Malenka, R. C. (1991) Agonists at metabotropic glutamate receptors presynaptically inhibit EPSPs in neonatal rat hippocampus. *J. Physiol.* **444,** 687–701.

Batchelor, A. M. and Garthwaite, J. (1993) Novel synaptic potentials in cerebellar Purkinje cells: probable mediation by metabotropic glutamate receptors. *Neuropharmacol.* **32,** 11–20.

Ben-Ari, Y., Aniksztejn, L., Otani, S., and Roisin, M. P. (1993) Quisqualate metabotropic glutamate receptors enhance NMDA currents and decrease the threshold for LTP induction through protein kinase C. *J. Neurosci.* **61,** S183.

Birse, E. F., Eaton, S. A., Jane, D. E., Jones, P. L. St. J., Porter, R. H. P., Pook, P. C.-K., Sunter, D. C., Udvarhelyi, P. M., Wharton, B., Roberts, P. J., Salt, T. E., and

Watkins, J. C. (1993) Phenylglycine derivatives as new pharmacological tools for investigating the role of metabotropic glutamate receptors in the central nervous system. *Neuroscience* **52**, 481–488.

Bleakman, D., Rusin, C., Chard, P., Glaum, S. R., and Miller, R (1992) Potentiation of ionotropic glutamate receptor signalling in substantia gelatinosa neurons by a metabotropic glutamate receptor agonist. *Mol. Pharm.* **42**, 192–196.

Boss, V. and Conn, P. J. (1993) Coupling of metabotropic excitatory amino acid receptors to phospholipase D: A novel pathway for generation of second messenger. *Funct. Neurol. Suppl.* **4**, 12.

Boss, V., Desai, M. A., Smith, T. S., and Conn, P. J. (1992) Trans-ACPD-induced phosphoinositide hydrolysis and modulation of hippocampal pyramidal cell excitability do not undergo parallel developmental regulation. *Brain Res.* **594**, 181–188.

Bossu, J. L., Fagni, L., Nooney, J., Bockaert, J., and Feltz, A. (1993) Increased Ca channel activity due to metabotropic glutamate receptor stimulation in isolated rat cerebellar granule cells. *J. Physiol.* **459**, 250P.

Caciagli, F., Casabona, G., L'Episcopo, M. R., Di Iorio, P., Ciccarelli, R., Shinozaki, H. and Nicoletti, F. (1993) Activation of metabotropic receptors reduces adenosine release in rat hippocampal slices. *Funct. Neurol. Suppl.* **4**, 13.

Cartmell, J., Kemp, J. A., Alexander, S. P. H., Hill, S. T., and Kendall, D. A. (1992) Inhibition of forskolin stimulated cyclic AMP formation by 1-aminocyclopentane-*trans*-1,3,dicarboxylate in guinea-pig cerebral cortical slices. *J. Neurochem.* **58**, 1964–1966.

Catania, M. V., Landwehrmeyer, B., Standaert, D., Testa, C., Penney, J. B., and Young, A. B. (1993) Differential expression patterns of metabotropic glutamate receptor mRNAs and binding sites in developing and adult rat brain. *Funct. Neurol. Suppl.* **4**, 15.

Charpak, S., Gähwiler, B. H., Do, K. Q., and Knöpfel, T. (1990) Potassium conductances in hippocampal neurons blocked by excitatory amino acid transmitters. *Nature* **347**, 765–767.

Coffey, E. T., Herrero, I., Sihra, T. S., and Nicholls, D. G. (1993) Metabotropic receptor activation of glutamate release is PKC mediated. *J. Neurosci.* **61**, S253.

Constanti, A. and Libri, V. (1992) *Trans*-ACPD induces a slow post-stimulus inward tail current (I_{ADP}) in guinea-pig olfactory cortex neurones in vitro. *Eur. J. Pharmacol.* **214**, 105,106.

Crepel, F., Daniel, H., Hemart, N., and Jaillard, D. (1991) Effects of ACPD and AP3 on parallel-fibre-mediated EPSPs of Purkinje cells in cerebellar slices in vitro. *Exp. Brain Res.* **86**, 402–406.

Davies, J. and Watkins, J. C. (1982) Actions of D and L forms of 2-amino-5-phosphonovalerate and 2-amino-4-phosphonobutyrate in the cat spinal cord. *Brain Res.* **235**, 378–386.

Desai, M. A. and Conn, P. J. (1991) Excitatory effects of ACPD receptor activation in the hippocampus are mediated by direct effects on pyramidal cells and blockade of synaptic inhibition. *J. Neurophys.* **66**, 40–52.

Gerber, U., Lüthi, A., and Gähwiler, B. H. (1993) Inhibition of a slow synaptic response by a metabotropic glutamate receptor antagonist in hippocampal CA3 pyramidal cells. *Proc. Royal Acad. Soc.* **254**, 169–172.

Glaum, S. R. and Miller, R. J. (1992) Metabotropic glutamate receptors mediate excitatory transmission in the nucleus of the solitary tract. *J. Neurosci.* **12,** 2251–2258.

Glaum, S. R. and Miller, R. J. (1993a) Activation of metabotropic glutamate receptors produces reciprocal regulation of ionotropic glutamate and GABA responses in the nucleus tractus solitarius. *J. Neurosci.* **13,** 1636–1641.

Glaum, S. R. and Miller, R. J. (1993b) Metabotropic glutamate receptors depress afferent excitatory transmission in the rat nucleus tractus solitarii. *J. Neurophys.* **70,** 2669–2672.

Glaum, S. R. and Miller, R. J. (1993c) Zinc protoporphyrin-IX blocks the effects of metabotropic glutamate receptor activation in the rat nucleus tractus solitarii. *Mol. Pharmacol.* **43,** 965–969.

Glaum, S. R. and Miller, R. J. (1994) Inhibition of phosphoprotein phosphatases block metabotropic glutamate receptor effects in the rat nucleus tractus solitarii. *Mol. Pharmacol.,* in press.

Glaum, S. R., Scholz, W. K., and Miller, R. J. (1991) Acute and long term glutamate mediated regulation of $[Ca^{2+}]_i$ in rat hippocampal pyramidal neurons *in vitro. JPET* **253,** 1293–1302.

Glaum, S. R., Slater, N. T., Rossi, D. J., and Miller, R. J. (1992) The role of metabotropic glutamate (ACPD) receptors at the parallel fiber-Purkinje cell synapse. *J. Neurophysiol.* **68,** 1453–1462.

Glaum, S. R., Sunter, D. C., Udvarhelyi, P. M., Watkins, J. C., and Miller, R. J. (1993) The actions of phenylglycine derived metabotropic glutamate receptor antagonists on multiple (1S,3R)-ACPD responses in the rat nucleus of the tractus solitarius. *Neuropharmacology* **32,** 1419–1425.

Greene, C., Schwindt, P., and Crill, W. (1992) Metabotropic receptor mediated after depolarization in neocortical neurons. *Eur. J. Pharmacol.* **226,** 279,280.

Harris, E. W. and Cotman, C. W. (1983) Effects of acidic amino acid antagonists on paired-pulse potentiation at the lateral perforant path. Exp. *Brain Res.* **52,** 455–460.

Hayashi, Y., Momiyama, A., Takahashi, T., Ohishi, H., Ogawa-Meguro, R., Shigemoto, R., Mizuno, N., and Nakanishi, S. (1993) Role of a metabotropic glutamate receptor, mGluR2, in synaptic modulation in the accessory olfactory bulb. *Nature* **366,** 687–690.

Holler, T., Klein, J., and Löffelholz, K. (1993) Glutamate activates phospholipase D in rat hippocampus. *Funct. Neurol. Suppl.* **4,** 26.

Hu, G.-Y. and Storm, J. F. (1991) Excitatory amino acids acting on metabotropic glutamate receptors broaden the action potential in hippocampal neurons. *Brain Res.* **568,** 339–344.

Irving, A. J., Schofield, G., Watkins, J. C., Sunter, D. C., and Collingridge, G. L. (1990) (1S,3R)-ACPD stimulates and L-AP3 blocks Ca^{2+} mobilization in rat cerebellar neurons. *Eur. J. Pharmacol.* **186,** 363–365.

Irving, A. J., Boulton, C. L., Garthwaite, J., and Collingridge, G. L. (1993) cGMP may mediate transient synaptic depression in rat hippocampal slices. *J. Physiol.* **108,** 89P.

Ishida, M., Saitoh, T., Shimamoto, K., Ohfune, Y., and Shinozaki, H. (1993) A novel metabotropic glutamate receptor agonist: Marked depression of monosynaptic excitation in the newborn rat isolated spinal cord. *Br. J. Pharmacol.* **109,** 1169–1177.

Ito, M and Karachot, L. (1990) Messengers mediating long-term desensitization in cerebellar Purkinje cells. *NeuroReport* **1,** 129–132.

Kahle, J. S. and Cotman, C. W. (1993) L-2-amino-4-phosphonobutanoic acid and 1S,3R-1-aminocyclopentane-1,3-dicarboxylic acid reduce paired-pulse depression recorded from medial perforant path in the dentate gyrus of rat hippocampal slices. *JPET* **266,** 207–215.

Katz, P. S. and Levitan, I. B. (1993) Quisqualate and ACPD are agonists for a glutamate-activated current in identified Aplysia neurons. *J. Neurophys.* **69,** 143–150.

Kelso, S. R., Nelson, T. E., and Leonard, J. P. (1992) Protein kinase C-mediated enhancement of NMDA currents by metabotropic glutamate receptors in xenopus oocytes. *J. Physiol.* **449,** 705–718.

Kinney, G. A. and Slater, N. T. (1993) Potentiation of NMDA receptor-mediated transmission in turtle cerebellar granule cells by activation of metabotropic glutamate receptors. *J. Neurophys.* **69,** 585–594.

Koerner, J. F. and Cotman, C. W. (1981) Micromolar L-2-amino-4-phosphonobutyric acid selectively inhibits perforant path synapses from entorhinal cortex. *Brain Res.* **216,** 192–198.

Konnerth, A., Llano, I., and Armstrong, C. M. (1990) Synaptic currents in cerebellar Purkinje cells. *Proc. Natl. Acad. Sci. USA* **57,** 2662–2665.

Kovalchuk, Y., Garaschuk, O., and Krishtal, O. A. (1993) Glutamate induces long-term increase in the frequency of single N-methyl-D-aspartate channel openings in hippocampal CA1 neurons examined in situ. *Neuroscience* **54,** 557–559.

Lanthorn, T. H., Ganong, A. H., and Cotman, C. W. (1984) 2-amino-4-phosphonobutyrate selectively blocks mossy fiber-CA3 responses in guinea pig but not rat hippocampus. *Brain Res.* **290,** 174–178.

Lester, R. A. and Jahr, C. E. (1990) Quisqualate receptor-mediated depression of calcium currents in hippocampal neurons. *Neuron* **4,** 741–749.

Linden, D. J. and Connor, J. A. (1991) Participation of postsynaptic PKC in cerebellar long-term depression in culture. *Science* **254,** 1656–1659.

Liu, Y.-B., Disterhof, J. F., and Slater, N. T. (1993) Activation of metabotropic glutamate receptors induces long-term depression of GABAergic inhibition in hippocampus. *J. Neurophys.* **69,** 1000–1004.

Llano, I., Dreessen, J., Kano, M., and Konnerth, A. (1991) Intradendritic release of calcium induced by glutamate in cerebellar Purkinje cells. *Neuron* **7,** 577–583.

Lovinger, D. M. (1991) *Trans*-1-amino-1,3-dicarboxylic acid (t-ACPD) decreases synaptic excitation in rat striatal slices through a presynaptic action. *Neurosci. Lett.* **129,** 17–21.

Maines, M. (1993) Carbon monoxide: An emerging regulator of cGMP in the brain. *Mol. Cell. Neurosci.* **4,** 389–397.

Manzoni, O., Prezeau, L., Rassendren, F. A., Sladeczek, F., Curry, K., and Bockaert, J. (1992) Both enantiomers of 1-aminocyclopentyl-1,3-dicarboxylate are full agonists of metabotropic glutamate receptors coupled to phospholipase. *Mol. Pharmacol.* **42,** 322–327.

Manzoni, O., Fagni, L., Pin, J-.P., Rassendren, F., Poulat, F., Sladeczek, F., and Bockaert, J. (1990) *(trans)*-1-amino-cyclopentyl-1,3-dicarboxylate stimulates quisqualate phosphoinositide-coupled receptors but not ionotropic glutamate receptors in striatal neurons and *Xenopus* oocytes. *Mol. Pharmacol.* **38,** 1–6.

Marks, G. S., Brien, J. F., Nakatsu, K., and McLaughlin, B. F. (1991) Does carbon monoxide have a physiological function. *Trends Pharmacol. Sci.* **12,** 185–188.

Martin, L. J., Blackstone, C. D., and Price, D. L. (1993) Cellular localization of metabotropic glutamate receptors in adult and developing brain. *Funct. Neurol. Suppl.* **4,** 34,35.

McCormick, D. A. and von Krosigk, M. (1992) Corticothalamic activation modulates thalamic firing through glutamate "metabotropic" receptors. *Proc. Natl. Acad. Sci. USA* **89,** 2774–2778.

McDonald, J. W. and Schoepp, D. D. (1993) The metabotropic excitatory amino acid receptor agonist (1S,3R)-ACPD selectively potentiates NMDA induced brain injury. *Eur. J. Pharmacol.* **215,** 353,354.

Miles, R. and Poncer, J.-C. (1993) Metabotropic glutamate receptors mediate a post-tetanic excitation of guinea-pig hippocampal inhibitory neurones. *J. Physiol.* **463,** 461–473.

Miller, R. J. (1990) Receptor-mediated regulation of calcium channels and neurotransmitter release. *FASEB J.* **4,** 3291–3299.

Milligan, G. (1993) Mechanisms of multifunctional signalling by G protein-linked receptors. *Trends Pharmacol. Sci.* **14,** 239–244.

Mori-Okamoto, J., Okamoto, K., and Tatsuno, J. (1993) Intracellular mechanisms underlying the suppression of AMPA responses by trans-ACPD in cultured chick purkinje neurons. *Mol. and Cell Neurosci.* **4,** 375–386.

Mulkey, R. M. and Zucker, R. S. (1992) Posttetanic potentiation at the crayfish neuromuscular junction is dependent on both intracellular calcium and sodium ion accumulation. *J. Neurosci.* **12,** 4327–4336.

Nadler, J. V., Zhou, M., and Duncan, C. P. (1993) Release of glutamate and aspartate from synaptosomes of the hippocampal schaffer collateral-commissural pathway. *J. Neurosci.* **61,** S253.

Nakajima, Y., Iwakabe, H., Akazawa, C., Nawa, H., Shigemoto, R., Mizuno, N., and Nakanishi, S. (1993) Molecular characterization of a novel retinal metabotropic glutamate receptor mGluR6 with a high agonist selectivity for L-2-amino-4-phosphonobutyrate. *J. Biol. Chem.* **268,** 11,868–11,873.

Nakanishi, S. (1992) Molecular diversity of glutamate receptors and implications for brain function. *Science* **258,** 597–603.

Nawy, S. and Jahr, C. E. (1991) cGMP-gated conductance in retinal bipolar cells is suppressed by the photoreceptor transmitter. *Neuron* **7,** 677–683.

Nicoletti, F., Meek, J. L., Iadorola, M. J., Chuang, D. M., Roth, B. L., and Costa, E. (1986) Coupling of inositol phospholipid metabolism with excitatory amino acid recognition sites rat hippocampus. *J. Neurochem.* **40,** 40–46.

Okada, D. (1992) Two pathways of cGMP production through glutamate receptor mediated nitric oxide synthesis. *J. Neurochem.* **59**, 1203–1210.

Pacelli, G. J. and Kelso, S. R. (1991) Trans-ACPD reduces multiple components of synaptic transmission in the rat hippocampus. *Neurosci. Lett.* **132**, 267–269.

Palmer, E., Monaghan, D. T., and Cotman, C. W. (1989) *Trans*-ACPD, a selective agonist of the phosphoinositide-coupled excitatory amino acid receptor. *Eur. J. Pharmacol.* **166**, 585–587.

Pook, P. C.-K., Sunter, D. C., Udvarhelyi, P. M., and Watkins, J. C. (1992) Evidence for presynaptic depression of monosynaptic excitation in neonatal rat moto-neurones by (1*S*,3*S*)- and (1*S*,3*R*)-ACPD. *Exp. Physiol.* **77**, 529–532.

Priddy, M., Drewe, J. A., and Kunze, D. L. (1992) L-glutamate inhibition of an inward potassium current in neonatal neurons from the nucleus of the solitary tract. *Neurosci. Lett.* **136**, 131–135.

Rainnie, D. G. and Shinnick-Gallagher, P. (1992) *Trans*-ACPD and L-APB presynaptically inhibit excitatory glutamatergic transmission in the basolateral amygdala (BLA). *Neurosci. Lett.* **139**, 87–91.

Randall, A. D., Wheeler, D. B., and Tsien, R. W. (1993) Modulation of Q-type Ca^{2+} channels and Q-type Ca^{2+} channel-mediated synaptic transmission by metabotropic and other G-protein linked receptors. *Funct. Neurol. Suppl.* **4**, 44,45.

Raymond, L. A., Blackstone, C. D., and Huganir, R. L. (1993) Phosphorylation and modulation of recombinant GluR6 glutamate receptors by cAMP-dependent protein kinase. *Nature* **361**, 637–641.

Riedel, G. and Reymann, K. (1993) An antagonist of the metabotropic glutamate receptor prevents LTP in the dentate gyrus of freely moving rats. *Neuropharmacol.* **9**, 929–931.

Ruzicka, B. B., and Jhamandas, K. H. (1993) Excitatory amino acid action on the release of brain neurotransmitters and neuromodulators: Biochemical studies. *Prog. Neurobiol.* **40**, 223–247.

Sacaan, A. I., Bymaster, F. P., and Schoepp, D. D. (1992) Metabotropic glutamate receptor activation produces extrapyramidal motor system activation that is mediated by striatal dopamine. *J. Neurochem.* **59**, 245–251.

Sahara, Y. and Westbrook, G. L. (1993) Modulation of calcium currents by a metabotropic glutamate receptor involves fast and slow kinetic components in cultured hippocampal neurons. *J. Neurosci.* **13**, 3041–3050.

Sanchez-Prieto, J., Herrero, J., and Miras-Portugal, M. T. (1993) Potentiation of glutamate exocytosis by a presynaptic glutamate metabotropic receptor. *J. Neurosci.* **61**, S253.

Schoepp, D. D. (1993) The biochemical pharmacology of metabotropic glutamate receptors. *Biochem. Soc. Trans.* **21**, 97–102.

Sladeczek, F., Pin, J.-P., Recasens, M., Bockaert, J., and Weiss, S. (1985) Glutamate stimulates inositol phosphate formation in striatal neurons. *Nature* **317**, 717–719.

Sladeczek, F., Momiyama, A., and Takahashi, T. (1993) Presynaptic inhibitory action of a metabotropic glutamate receptor agonist on excitatory transmission in visual cortical neurons. *Funct. Neurol. Suppl.* **4**, 52.

Staub, C., Vranesic, I., and Knöpfel, T. (1992) Responses to metabotropic glutamate receptor activation of cerebellar Purkinje cells: induction of an inward current. *Eur. J. Neurosci.* **4**, 832–839.

Stelzer, A. and Wong R. K. S. (1989) GABA-A responses in hippocampal neurons are potentiated by glutamate. *Nature* **337**, 170–173.

Stevens, C. F. and Wang, Y. (1993) Reversal of long-term potentiation by inhibitors of haem oxygenase. *Nature* **364**, 147–149.

Stratton, K. R., Worley, P. F., and Baraban, J. M. (1989) Excitation of hippocampal neurons by stimulation of glutamate Q_p receptors. *Eur. J. Pharmacol.* **173**, 531–533.

Sugiyama, H., Ito, I., and Watanabe M. (1989) Glutamate receptor subtypes may be classified into two major categories: A study on *Xenopus* oocytes injected with rat brain mRNA. *Neuron* **3**, 129–132.

Swartz, K. (1993) Modulation of Ca^{2+} channels by protein kinase C in rat central and peripheral neurons: Disruption of G protein-mediated inhibition. *Neuron* **11**, 305–320.

Swartz, K. J. and Bean, B. P. (1992) Inhibition of calcium channels in rat CA3 pyramidal neurons by a metabotropic glutamate receptor. *J. Neurosci.* **12**, 4358–4371.

Swartz, K. J., Merrit, A., Bean, B. P., and Lovinger, D. M. (1993) Protein kinase C modulates glutamate receptor inhibition of Ca^{2+} channels and synaptic transmission. *Nature* **361**, 165–168.

Tanabe, Y., Nomura, A., Masu, M., Shigemoto, R., Mizuno, N., and Nakanishi, S. (1993) Signal transduction, pharmacological properties, and expression patterns of two rat metabotropic glutamate receptors, mGluR3 and mGluR4. *J. Neurosci.* **13**, 1372–1378.

Thomsen, C., Kristensen, P., Mulvihill, E., Haldeman, B., and Suzdak, P. D. (1992) L-AP4 is an agonist at the type IV metabotropic glutamate receptor which is negatively coupled to adenylyl cyclase. *Eur. J. Pharmacol. Mol. Pharmacol.* **227**, 361–362.

Verma, A., Hirsch, D. J., Glatt, C. E., Ronnett, G. V., and Snyder, S. H. (1993) Carbon monoxide: A putative neural messenger. *Science* **259**, 381–384.

Vranesic, I., Batchelor, A., Gähwiler, B. H., Garthwaite, J., Staub, C., and Knöpfel, T. (1991) *Trans*-ACPD-induced Ca^{2+} signals in cerebellar Purkinje cells. *Neuroreport* **2**, 759–762.

Vranesic, I., Staub, C., and Knöpfel, T. (1993) Activation of metabotropic glutamate receptors induces an outward current which is potentiated by methylxanthines in rat cerebellar Purkinje cells. *Neurosci. Res.* **16**, 209–215.

Westbrook, G. L., Sahara, Y., Saugstad, J. A., Kinzie, J. M., and Segerson, T. P. (1993) Regulation of ion channels by ACPD and AP4. *Funct. Neurol. Suppl.* **4**, 56.

Winder, D. G. and Conn, P. J. (1993) Activation of metabotropic glutamate receptors increases cAMP accumulation in hippocampus by potentiating responses to endogenous adenosine. *J. Neurosci.* **13**, 38–44.

Yool, A. J., Krieger, R. M., and Gruol, D. L. (1992) Multiple ionic mechanisms are activated by the potent agonist quisqualate in cultured cerebellar purkinje neurons. *Brain Res.* **573**, 83–94.

Zeilhofer, H. U., Muller, T. H., and Swandulla, D. (1993) Inhibition of high voltage-activated calcium currents by L-glutamate receptor-mediated calcium influx. *Neuron* **10,** 879–887.

Zheng, F. and Gallagher, J. P. (1992) Burst firing of rat septal neurons induced by (1S,3R)-ACPD requires influx of extracellular calcium. *Eur. J. Pharmacol.* **211,** 281–282.

Zhu, P. J. and Krnjevic, K. (1993) Adenosine release is a major cause of failure of synaptic transmission during hypoglycemia in rat hippocampal slices. *Neurosci. Lett.* **155,** 128–131.

Chapter 7

Long-Lasting Modulation
of Synaptic Transmission
by Metabotropic Glutamate Receptors

Joel P. Gallagher, Fang Zheng,
and Patricia Shinnick-Gallagher

1. Introduction

Studies of the functional roles for metabotropic glutamate receptors (mGluRs) in the central nervous system have rapidly developed over the past two years. mGluR agonists have a diverse range of electrophysiological effects, including long-lasting modulation of synaptic transmission. Putative mGluR antagonists L-2-amino-4-phosphonobutyrate (L-AP4) and/or 2-amino-3-phosphonopropionic acid (AP3) blocked the induction or/and maintenance of long-term potentiation (LTP) in the hippocampal CA1 region (Reymann and Matthies, 1989; Izumi et al., 1991) and the dorsolateral septal nucleus (DLSN) of rat (Zheng and Gallagher, 1990). Phospholipase C (PLC)-coupled mGluRs may also play a role in the induction of long-term depression in cerebellar Purkinje neu-

The Metabotropic Glutamate Receptors Eds.: P. J. Conn and J. Patel
© 1994 Humana Press Inc., Totowa, NJ

rons (Linden et al., 1991; Daniel et al., 1992). In addition to the long-lasting modulation of synaptic transmission, mGluRs might also play a role in the pathogenesis of epilepsy. In this chapter, we will review the current data and discuss the problems encountered in this field.

2. Roles of mGluRs in Long-Term Potentiation

LTP is a sustained increase of synaptic efficacy induced by high-frequency electrical stimulation of selected pathways in mammalian brain (Bliss and Lomo, 1973), and is currently the most compelling cellular model for learning and memory (Morris et al., 1986). The induction of LTP requires an influx of extracellular calcium into postsynaptic neurons through either channels gated by N-methyl-D-aspartate (NMDA) receptors during tetanus (Collingridge et al., 1983; Harris et al., 1984; Artola and Singer, 1987; Schmidt, 1990) or an influx of extracellular calcium through voltage-dependent Ca^{2+} channels (Grover and Teyler, 1990). However, the rise of intracellular calcium concentration that is essential for the induction of LTP (Lynch et al., 1983) could also theoretically be achieved by releasing calcium from intracellular calcium storage sites, such as the inositol trisphosphate (IP_3)-sensitive internal store, via the activation of a PLC-coupled mGluR (Sladeczek et al., 1985; Murphy and Miller, 1988). Early reports supported a possible role for mGluRs in induction and maintenance of LTP in both hippocampus (Reymann and Matthies, 1989) and dorsolateral septal nucleus (DLSN) (Zheng and Gallagher, 1990). Recent studies have provided further evidence to support such a role of mGluRs at mossy fiber-CA3 pyramidal neuron synapses (Ito and Sugiyama, 1991), Schaffer collateral-CA1 pyramidal neuron synapses (McGuinness et al., 1991b; Otani and Ben-Ari, 1991; Radpour and Thomson, 1992; Bortolotto and Collingridge, 1992; Behnisch and Reymann, 1993; Bashir et al., 1993a), and fimbrial fiber-DLSN synapses (Zheng and Gallagher, 1992b). However, lack of selective pharmacological tools (*see* Chapter 4), the explosion of information regarding the identities of multiple mGluRs (*see* Chapter 1), and knowledge of a differential distribution of mGluRs as determined by mRNA (Table 1; *see* Chapter 4) have left certain questions unanswered.

Table 1
Distribution of mRNAs in Major Brain Structures of Rat

Brain regions	mGluR1[a]	mGluR2[b]	mGluR3[c]	mGluR4[d]	mGluR5[e]
Cerebral cortex	+	++	++	−	+++
Cerebellar cortex					
Purkinje cells	++++	−	−	−	−
Granule cells	+	−	−	++++	−
Golgi cells	++	++++	++	−	+
Hippocampus					
CA1 pyramidal cells	+	−	−	−	+++
CA3 pyramidal cells	+++	−	−	+	+++
Dentate gyrus					
Granule cells	++	++	+	+	+++
CA4 pyramidal cells	+++	−	−	+	+++
Septum					
Dorsolateral	+++	−	−	+	+++
Medial	−	+/−	−		
Basal ganglia					
Striatum	++	+	+		+++
Nucleus accumbens	+	+	+		+++
Amygdala					
Basolateral	+	+++	++		
Medial	+	+++	−		
Central	+				
Main olfactory bulb					
Mitral cells	+++	−	−		
Tufted cells	+++	−	−		
Interior granule cells	+	+	−	+++	+++
Accessory olfactory bulb					
Mitral cells	+++	+++	−		
Granule cells	+	++	−		
Thalamus				+++	
Anterior nucleus	++	++	−		+++
Reticular nucleus			+++		

++++, Very high; +++, High; ++, Moderate; +, Low; −, Background Level.
[a]Shigemoto, R., Nakanishi, S., and Mizuno, N. (1992) *J. Comp. Neurol.* **322**, 121–135.
[b]Ohishi et al. (1993) *Neuroscience* **53**, 1009–1018.
[c]Ohishi et al. (1993) *J. Comp. Neurol.* **335**, 252–266.
[d]Tanabe et al. (1993) *J. Neurosci.* **13**, 1372–1378.
[e]Abe et al. (1992) *J. Biol. Chem.* **267**, 13,361–13,368.

2.1. Long-Term Potentiation at the Mossy Fiber Synapses in the Hippocampal CA3 Region

The exact locus for the induction of mossy-fiber LTP is still controversial, however, it is undoubtedly clear that the mossy fiber LTP is NMDA receptor-independent (Harris and Cotman, 1986; Williams and Johnston, 1988; Zalutsky and Nicoll, 1990). One possible mechanism for the induction of mossy fiber LTP is that calcium enters postsynaptic neurons through voltage-gated calcium channels. This hypothesis is supported by several lines of evidence, which have been reviewed previously (Johnston et al., 1992). However, one report suggested that the induction of mossy fiber LTP is presynaptic (Zalutsky and Nicoll, 1992).

Several lines of evidence suggest that mGluRs might be involved in the induction of mossy fiber LTP. First, mossy fiber LTP is blocked by intracerebroventricular injection of pertussis toxin (PTX) (Ito et al., 1988). Therefore, a PTX-sensitive G-protein is required for the induction of LTP. This PTX-sensitive G-protein seems to be located at presynaptic terminals, since intracellular perfusion of GTPγS blocked LTP of fimbrial fiber-evoked EPSPs, but failed to block mossy fiber LTP in CA3 pyramidal neurons (Katsuki et al., 1992). Second, D,L-AP3 at 1 mM reduced the mossy fiber LTP (Ito and Sugiyama, 1991). Finally, ibotenate elicited a long-lasting enhancement of mossy fiber EPSPs, which was also blocked by D,L-AP3 (Ito and Sugiyama, 1991). However, these data are not conclusive.

Hippocampal CA3 pyramidal neurons (Table 1) have a high level of mRNAs for two PLC-coupled mGluRs, i.e., mGluR1 and mGluR5 (Shigemoto et al., 1992; Abe et al., 1992). However, these receptors are unlikely to play any role in the mossy fiber LTP, since the PTX-sensitive "metabotropic" receptor involved has to be presynaptic, because it is not affected by intracellular injection of GTPγS (Katsuki et al., 1992). Dentate granule cells also have moderate to high levels of mRNAs for mGluR1 and mGluR5. If these receptors are involved in the mossy fiber LTP, quisqualate, the most potent agonist for these receptors, should mimic ibotenate to induce a long-lasting potentiation of EPSPs at mossy fiber synapses. The fact that quisqualate failed to induce mossy fiber LTP (Ito and Sugiyama, 1991) strongly suggests that ibotenate-induced mossy fiber LTP is not mediated by mGluR1 or mGluR5. Furthermore, D,L-AP3 is not an effective antagonist for these receptors (for review, *see* Chapter 4). At 1 mM, D,L-AP3 will act preferentially as an agonist at mGluRs negatively coupled to the cAMP cascade. A more reasonable interpretation of the effects of D,L-AP3 on mossy fiber LTP would be that activation of mGluRs coupled negatively to the cAMP cascade might have effects similar to those of activation of muscarinic receptors (Williams and Johnston, 1988), i.e., to block the induction of mossy fiber LTP.

2.2. Long-Term Potentiation at the Schaffer Collateral Synapses in the Hippocampal CA1 Region

There have been a lot of conflicting data about the role of mGluRs in the induction and maintenance of Schaffer collateral LTP. The initial controversy was whether the induction of Schaffer collateral LTP is sensitive to PTX. An earlier report (Ito et al., 1988) suggested that Schaffer collateral LTP induced by tetanic stimuli is not blocked by PTX injected into lateral ventricles. However, a later report with an improved injection method (Goh and Pennefather, 1989) demonstrated that PTX blocked the induction of Schaffer collateral LTP. Yet, intracellular injection of GTPγS into hippocampal CA1 pyramidal neurons failed to block the induction of Schaffer collateral LTP (Goh and Pennefather, 1989; Katsuki et al., 1992). Therefore, the PTX-sensitive G-protein is most likely located at a presynaptic, rather than a postsynaptic site.

The effects of AP4 and AP3, which were once thought to be antagonists at PLC-coupled mGluRs, have been studied on Schaffer collateral LTP by several groups. They obtained data that conflicted in many aspects. One group (Izumi et al., 1991) reported that AP3 applied before or immediately after tetanic stimuli blocked Schaffer collateral LTP, suggesting that AP3 blocked both the induction and maintenance phase of LTP. However, two other groups (Stanton et al., 1991; Behnisch and Reymann, 1993) reported that AP3 (including the L-isomer) failed to block the induction of Schaffer collateral LTP. All three groups were measuring the slope of field EPSPs and the amplitude of the population spikes, while recording from adult rats; thus, there are no apparent technical differences that could help to explain the conflicting results. On the other hand, both L-AP3 and L-AP4 blocked the late phase of Schaffer collateral LTP, i.e., that phase of LTP observed later than 60 min after tetanic stimuli (Reymann and Matthies, 1989; Behnisch and Reymann, 1993). Although D-AP3 (300 μM) was ineffective (Behnisch and Reymann, 1993), D-AP4 was almost as potent as L-AP4 (Reymann and Matthies, 1989). Since the phosphoinositide hydrolysis in the adult hippocampus is not consistently blocked by L-AP3 (Schoepp and Johnson, 1989; Vecil et al., 1992) and is not blocked at all by D-AP4 (Schoepp and Johnson, 1988), it is unlikely that the inhibition of late LTP is the result of blockade of PLC-coupled mGluRs in the hippocampal CA1 pyramidal neurons.

The selective agonist *trans*-(±)-1-amino-1,3-cyclopentanedicarboxylic acid (*trans*-ACPD) or its active enantiomer 1S,3R-ACPD has been shown consistently to potentiate the tetanic stimuli-induced Schaffer collateral LTP (McGuinness et al., 1991a,b; Otani and Ben-Ari, 1991; Behnisch and Reymann, 1993; Otani et al., 1993). There is a consensus that the potentiation of LTP by *trans*-ACPD or 1S,3R-ACPD is because of an acute and reversible potentiation of NMDA receptors (Aniksztejn et al., 1992; Harvey and Collingridge, 1993). However, the second-messenger pathways involved in this potentiation have been disputed. Two groups reported that the potentiation of an NMDA current and LTP is mediated by activation of PKC (Aniksztejn et al., 1992; McGuinness et al., 1991a), because it was blocked by 10 μM sphingosine (McGuinness et al., 1991a; Aniksztejn et al., 1992) or a combination of sphingosine and PKC (19–36) peptide inside the electrode (Aniksztejn et al., 1992). These observations are contradicted by a recent report (Harvey and Collingridge, 1993) that the potentiation of a NMDA current by 1S,3R-ACPD involves a calcium-dependent process and is negatively regulated by PKC.

1S,3R-ACPD or *trans*-ACPD alone failed to induce any long-lasting potentiation of Schaffer collateral EPSPs in most reports

(Baskys and Malenka, 1991; Desai and Conn, 1991; Otani and Ben-Ari, 1991; McGuinness et al., 1991a; Radpour and Thomson, 1992; Behnisch and Reymann, 1993; Collins and Davies, 1993) with one exception (Bortolotto and Collingridge, 1992, 1993). In this latter study (Bortolotto and Collingridge, 1993), superfusion of 10 μM 1S,3R-ACPD induced a slowly developing LTP. This ACPD-induced LTP is not blocked by AP5, but is blocked by broad-spectrum protein kinase inhibitors and thapsigargin (Bortolotto and Collingridge, 1993). This slow-developing LTP induced by 1S,3R-ACPD is similar to ACPD-induced LTP observed in the rat DLSN (Zheng and Gallagher, 1992b). It is uncertain why this potentiation is not observed by other groups using similar or higher concentrations of 1S,3R-ACPD or *trans*-ACPD. One group reported that coactivation of *trans*-ACPD and NMDA resulted in a slow-developing LTP (Radpour and Thomson, 1992). This observation could not be replicated by Behnisch and Reymann (1993). Furthermore, another group reported that *trans*-ACPD, in combination with arachidonic acid, potentiates synaptic transmission, whereas *trans*-ACPD or arachidonic acid applied alone caused inhibition of synaptic transmission (Collins and Davies, 1993).

1S,3R-ACPD has been suggested to induce the slow-developing LTP by activation of a PLC-coupled mGluR (Bortolotto and Collingridge, 1993). However, 10 μM 1S,3R-ACPD does not significantly stimulate PI hydrolysis in adult hippocampal slices (Schoepp et al., 1991). Furthermore, *trans*-ACPD is a weak agonist at the two PLC-coupled cloned mGluRs with EC_{50} values above 100 μM (Masu et al., 1991; Abe et al., 1992). On the other hand, *trans*-ACPD is a very potent agonist at the two mGluRs coupled negatively to the cAMP cascade, i.e., mGluR2 and mGluR3, with EC_{50} values ranging from 5 to 8 μM (Tanabe et al., 1992, 1993). In adult hippocampus, 1S,3R-ACPD inhibited the forskolin-stimulated cAMP formation with an estimated EC_{50} of 10 μM (Schoepp et al., 1992). Responses mediated by both mGluR2 and mGluR3 are sensitive to PTX (Tanabe et al., 1992; Tanabe et al., 1993). As discussed earlier, a presynaptic PTX-sensitive receptor is required for the induction of Schaffer collateral LTP. Thus, it is possible that the ACPD-induced LTP is the result of activation of a presynaptic mGluR coupled negatively to the cAMP cascade.

Postsynaptic PLC-coupled mGluRs may play a role in the tetanic stimuli-induced Schaffer collateral LTP. A phenylglycine derivative, α-Methyl-4-carboxyphenylglycine (MCPG), has been recently suggested to be an effective mGluR1 antagonist (Bashir et al., 1993a). This drug blocked the tetanic stimuli-induced Schaffer collateral LTP in the hippocampal CA1 region (Bashir et al., 1993a) and reduced a tetanic stimuli-induced rise of dendritic calcium levels (Frenguelli et al., 1993). These observations have not proven conclusively that postsynaptic PLC-coupled mGluRs are an absolute requirement for the induction of tetanic stimuli-induced Schaffer collateral LTP, since MCPG also blocks other mGluRs. Intracellular injection of a monoclonal antiphosphatidylinositol-4,5-*bis*-phosphate (PIP$_2$) antibody also blocked the induction of Schaffer collateral LTP (Tsubokawa et al., 1993). However, this result could be interpreted in many ways.

2.3. Long-Term Potentiation in the Dentate Gyrus

The late phase of LTP in the dentate gyrus of freely moving rats was blocked by MCPG, a putative mGluR antagonist (Riedel and Reymann, 1993). Furthermore, a NMDA receptor-mediated EPSC was potentiated by both tetanic stimuli and 10 μM 1S,3R-ACPD (O'Connor et al., 1993b). The LTP of NMDA receptor-mediated EPSCs by tetanic stimuli or 1S,3R-ACPD was reduced by 200 μM L-AP3 or 400 μM MCPG (O'Connor et al., 1993a). More dramatically, the duration of the potentiation was greatly reduced (O'Connor et al., 1993a). These observations suggest a possible role for mGluRs in the maintenance of the non-NMDA and NMDA components of EPSPs in the dentate gyrus.

2.4. Long-Term Potentiation
at the Rat Dorsolateral Septal Nucleus

At rat dorsolateral septal nucleus synapses, tetanic stimuli-induced LTP is independent of the NMDA receptor (Zheng and Gallagher, 1992b). The induction of LTP was blocked by L-AP4 (100–200 μM) or L-AP3 (50 μM), whereas D-AP4 was ineffective

(Zheng and Gallagher, 1992b). L-AP3 blocked the induction, but not the maintenance of tetanic stimuli-induced LTP because L-AP3 applied immediately after tetanic stimuli failed to block the induction of LTP at rat dorsolateral septal nucleus synapses (Zheng and Gallagher, 1992b). The induction of this NMDA-independent LTP was a postsynaptic process because it was blocked by intracellular injection of GTPγS (Zheng and Gallagher, 1992b). The requirement of a postsynaptic G-protein strongly suggests that the dorsolateral septal nucleus LTP is mGluR-dependent.

Superfusion of 20 μM 1S,3R-ACPD induced a slow-developing LTP similar to that observed in the hippocampal CA1 region. This ACPD-induced LTP was also blocked by L-AP3 and GTPγS, suggesting that it was mediated by a mGluR (Zheng and Gallagher, 1992b).

The tetanic stimuli-induced LTP at the dorsolateral septal nucleus was blocked by intracellular injection of BAPTA, a calcium chelator, indicating that a rise in intracellular calcium is the triggering signal for the induction of mGluR-dependent LTP (Zheng and Gallagher, 1992b). The tetanic stimuli-induced LTP was also blocked by thapsigargin, suggesting that the induction of mGluR-dependent LTP requires calcium release from internal stores (Zheng and Gallagher, 1992a). Therefore, we had speculated that the mGluRs involved in the dorsolateral septal nucleus LTP were coupled to PLC.

In situ hybridization analysis has demonstrated that the rat dorsolateral septal nucleus has high levels of mRNAs for both mGluR1 and mGluR5 (Table 1). Our biochemical study (Zheng et al., 1994) demonstrated that 1S,3R-ACPD raised the IP$_3$ level in the dorsolateral septal nucleus slices. However, the increase in IP$_3$ level required at least 30 μM 1S,3R-ACPD. Furthermore, L-AP3 failed to block the ACPD-stimulated IP$_3$ increase (Zheng et al., 1993). These observations are in agreement with the pharmacological data for mGluR1 and mGluR5 expressed in *Xenopus* oocytes or Chinese hamster ovary cells. The ineffectiveness of L-AP3 to block ACPD-stimulated PI hydrolysis suggests that the blockade of LTP by L-AP3 at the dorsolateral septal nucleus is not the result of blockade of either mGluR1 or mGluR5.

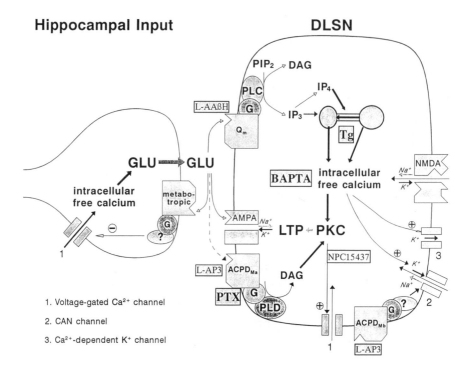

Hippocampal Input

DLSN

1. Voltage-gated Ca²⁺ channel
2. CAN channel
3. Ca²⁺-dependent K⁺ channel

Fig. 1. Model for the role of mGluRs at the dorsolateral septal nucleus synapse. Postsynaptic $ACPD_{Ma}$ is only activated during tetanic stimulation, whereas Q_m may be activated by low-frequency stimuli. On activation, Q_m elevates the intracellular free-calcium level by causing calcium release from IP_3-sensitive internal stores. $ACPD_{Ma}$ may further increase intracellular free-calcium level by potentiating a voltage-gated calcium current, resulting in an enhanced influx of extracellular calcium. $ACPD_{Mb}$, a PTX-resistant metabotropic receptor, may contribute to the increase of intracellular free-calcium level by potentiating Ca^{2+}-activated non-specific (CAN) channels in some dorsolateral septal nucleus neurons. The increase in intracellular free-calcium level in combination with activation of PKC directly by $ACPD_{Ma}$ triggers the induction of LTP. On the other hand, a presynaptic mGluR may be present to inhibit synaptic transmission at the dorsolateral septal nucleus. The presynaptic mGluR is pharmacologically different from the postsynaptic mGluRs.

Our recent pharmacological studies have indicated that there are multiple types of mGluRs in the rat dorsolateral septal nucleus (Zheng and Gallagher, 1993; Gallagher and Zheng, 1993; Fig. 1). 1S,3R-APCD induced a burst firing that was selectively blocked by PTX. Thus, the burst firing appears to be mediated by a distinct PTX-sensitive mGluR (Zheng and Gallagher, 1993). This receptor, tentatively named $ACPD_{Ma}$, is blocked by L-AP3 (50 μM) (Zheng et al., 1993), suggesting that it is not likely to be linked to PI turnover. We suggest that the $ACPD_{Ma}$ receptor is required for the induction of LTP at dorsolateral septal nucleus synapses. This hypothesis is supported by our data that PTX blocked the tetanic stimuli-induced LTP in the majority of dorsolateral septal nucleus neurons (manuscript submitted). The $ACPD_{Ma}$ receptor is likely linked to PKC via a PLC-independent mechanism, possibly the phospholipase D cascade (manuscript submitted). Boss and Conn (1992) have demonstrated that 1S,3R-ACPD activated the PLD cascade in hippocampal slices. However, our data with BAPTA and thapsigargin suggest that a mGluR coupled to PLC may also be required for the induction of LTP at the rat dorsolateral septal nucleus.

3. Roles of mGluRs in Long-Term Depression

mGluRs are also involved in the induction of long-term depression (LTD) in cerebellar Purkinje cells (Linden et al., 1991; Daniel et al., 1992), hippocampal pyramidal neurons (Stanton et al., 1991; Bashir et al., 1993b), striatal neurons (Calabresi et al., 1992), and visual cortex (Kato, 1993). In the cerebellum, LTD of the parallel-fiber Purkinje-neuron synapses was observed after coactivation of parallel fibers and climbing fibers. Conjunctive application of QA with climbing fiber stimulation also induced LTD of the parallel fiber Purkinje neuron synapses (Kano and Kato, 1987). Similar long-term depression of glutamate currents was induced by concurrent iontophoretic glutamate application and depolarization of the Purkinje neurons (Linden et al., 1991). When either aspartate, kainate or AMPA were combined with depolarization of the Purkinje neuron, no induction of LTD was observed; however, when a combination of AMPA and *trans*-ACPD was combined with depolarization of the Purkinje neuron, LTD was successfully induced (Linden et al., 1991). Furthermore, the LTD induced by glutamate/depolarization conjunction was blocked by both PTX and L-AP3 (Linden et al., 1991). Thus, it has been suggested that AMPA receptors and mGluRs must be coactivated in order to evoke LTD at the parallel fiber-

Purkinje neuron synapses. However, a recent study suggested that coactivation of mGluRs and voltage-gated calcium channels was sufficient to induce LTD at the parallel fiber-Purkinje neuron synapses, and that activation of AMPA receptors was not necessary (Daniel et al., 1992). In the hippocampal CA1 region, induction of LTD was blocked by L-AP3 (Stanton et al., 1991) or MCPG (Bashir et al., 1993b). At rat striatum, coactivation of mGluRs with dopamine receptors induced LTD (Calabresi et al., 1992). In rat visual cortex, the LTD induced by tetanic stimuli was blocked by intracellular injection of GTPγS, heparin or EGTA, a calcium chelator (Kato, 1993). Furthermore, the LTD in the visual cortex was induced without tetanic stimuli by superfusion of 10 μM quisqualate, but was not induced by 10 μM *trans*-ACPD (Kato, 1993). Thus, based on Kato's results, it appears that a quisqualate-sensitive mGluR may be involved in the induction of LTD in the visual cortex.

4. mGluRs in Kindling

Kindling is a long-lasting form of synaptic plasticity and is considered an animal model of human complex partial (temporal lobe) seizures (McNamara et al., 1989; Sato et al., 1990). Kindling results from repeated administration of subconvulsive electrical stimuli applied most often to the amygdala or to afferent projections to the hippocampus, which progressively develops into intense and generalized seizures (Goddard et al., 1969). This alteration in neuronal excitability is considered to be permanent (Cain, 1989) and is postulated to represent permanently established LTP (Douglas and Goddard, 1975; Goddard et al., 1978). The different stages of the development of a kindled seizure are usually graded on a five-point scale described by Racine (1972), stage 5 indicating a fully kindled seizure.

Glutamate receptors perhaps of the metabotropic type were first implicated in the induction of kindling about 10 years ago. Savage et al., (1984) showed that the number of quisqualate-sensitive glutamate binding sites on hippocampal membranes was increased selectively after three class 4 seizures elicited by stimulating the angular bundle.

These changes could not be detected 28 d after the last kindled seizure, suggesting these binding sites could be involved in the induction but not maintenance of the kindled state (Savage et al., 1984).

The first clear studies suggesting the involvement of mGluRs in kindling involved measurement of inositol phospholipid hydrolysis elicited by ibotenic acid (Iadorola et al., 1986). Kindling in the amygdala enhanced the stimulation of PI turnover induced by ibotenate at a stage 3–4 seizure. These effects returned to normal 1 mo after the cessation of electrical stimulation and were antagonized by 2-amino-4-phosphonobutyric acid. PI turnover elicited by carbachol (carbamylcholine) or norepinephrine was unaffected by kindling (Iadorola et al., 1986). These data suggested that the enhancement of PI turnover associated with kindling was specific to mGluRs and not the result of a generalized increase in PLC activity (Iadorola et al., 1986).

Further in-depth studies of ibotenate-stimulated PI turnover were carried out by Akiyama et al., (1987, 1989, 1992). This group showed that PI turnover was increased in the amygdala as well as the hippocampus of amygdala-kindled animals. The increase in the ibotenate-stimulated PI hydrolysis was observed in the amygdala/pyriform cortex 24 h and persisted 7 d after the last stage 5 kindled seizure; in contrast, the increase in the hippocampus was transient and observed only 24 h after the last seizure (Akiyama et al., 1987). Additional studies examined whether this increase was longer lasting (Akiyama et al., 1989). Four weeks after the last kindled seizure elicited by stimulating the amygdala, significant increases in ibotenate-stimulated PI turnover were recorded in the ipsilateral and contralateral sides of the amygdala/pyriform cortex. When the hippocampus was used to elicit kindling, ibotenate-stimulated PI turnover remained significantly elevated for 15 d in the hippocampus, but not 30 d after the last seizure (Yamada et al., 1989).

Finally, the long-lasting enhancement of PI turnover in the amygdala/pyriform cortex was recorded in response to a specific metabotropic agonist, ACPD. *Trans-* but not *cis-*ACPD elicited increases in PI turnover, which could be observed 28 d after kindling in the deep prepiriform cortex (Akiyama et al., 1992). These data suggest that mGluRs contribute to epileptogenesis particularly in the amygdala/pyriform cortex.

Using the amygdala kindling model of epilepsy in our laboratory, we have described long-term changes in neuronal excitability with epileptogenesis (Gean et al., 1989; Rainnie et al., 1992). Kindling significantly increased the amplitude of slow NMDA and fast-non-NMDA components of glutamatergic synaptic transmission in the basolateral amygdala. Furthermore, maximal excitatory postsynaptic potentials in kindled neurons were elicited at stimulus intensities significantly lower than those in control animals (Rainnie et al., 1992). These changes were recorded in amygdala slices contralateral to the site of stimulation up to 6 wk after the last kindled seizure in vivo.

We have recently examined the effects of mGluR activation on amygdala neurons in control and kindled animals. In control animals, 1S,3R-ACPD induces a membrane hyperpolarization accompanied by an increase in conductance; in some neurons, the hyperpolarization is followed by a depolarization (Rainnie and Shinnick-Gallagher, 1992; Rainnie et al., in press; Fig. 2A). Preliminary data show that the membrane hyperpolarization induced by 1S,3R-ACPD is not observed in kindled neurons (Fig. 2B; Rainnie, Holmes, and Shinnick-Gallagher, unpublished observations). These data suggest that electrophysiological responses mediated by mGluRs are profoundly changed in kindled animals.

The physiological role of mGluRs in epilepsy is unknown. The endogenous function of these receptors could not be tested in behavioral studies of epilepsy because of a lack of specific antagonists. In one study, 1S,3R-ACPD (1 μmol/2 μL) was injected into the hippocampus of halothane-anesthetized rats (Sacaan and Schoepp, 1992). The delayed-onset seizures that resulted resembled those produced by kindling an animal electrically to a stage 4 or stage 5 seizure. Seizures were prevented or suppressed by NMDA receptor and mGluR antagonists. Accompanying the seizures was neuronal damage similar to that produced by NMDA. However, higher concentrations of NMDA (10 μM) and 1R,3S-ACPD (3 mM), which can both directly activate NMDA receptors, were less potent in vivo in inducing neuronal damage (Sacaan and Schoepp, 1992). These data

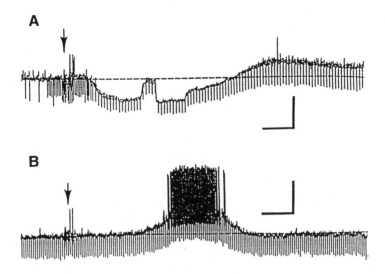

Fig. 2. Response to *trans*-ACPD in control and kindled neurons of the basolateral amygdala. **(A)** Membrane hyperpolarization induced by drop application of 1S,3R-ACPD (arrow; equivalent to 40 μM bath concentration) accompanied by an increase in conductance. During the hyperpolarization, the membrane was returned to control levels to ensure that the conductance change was not due to membrane rectification. A membrane depolarization associated with a decrease in conductance followed the membrane hyperpolarization. Resting membrane potential = –63 mV. **(B)** In a neuron from a kindled animal, drop application of *trans*-ACPD (arrow) cause a delayed membrane depolarization. The membrane conductance was decreased in individual electrotonic potentials although obscured by cell firing in the record. Resting membrane potential = –61 mV. Downward deflections represent electrotonic potentials generated in response to injecting 200-ms, 0.1-nA current pulses across the membrane as a measure of membrane resistance. Calibration: 10 mV × 20 s.

suggest that mGluRs may be involved in epileptogenesis and neuronal damage.

In summary, there are consistent and reproducible findings of a persistent enhancement of phosphoinositide hydrolysis mediated through mGluRs in kindled animals, suggesting a role for mGluRs in kindling. Preliminary electrophysiological and behavioral data also provide evidence for mGluR involvement in the etiology of seizures.

Acknowledgments

This work was supported by NS 24643 to P. S. G., MH 39163 and The Council For Tobacco Research, USA, Inc. to J. P. G.

References

Abe, T., Sugihara, H., Nawa, H., Shigemoto, R., Mizuno, N., and Nakanishi, S. (1992) Molecular characterization of a novel metabotropic glutamate receptor mGluR5 coupled to inositol phosphate/Ca^{2+} signal transduction. *J. Biol. Chem.* **267,** 13,361–13,368.

Akiyama, K., Yamada, N., and Otsuki, S. (1989) Lasting increase in excitatory amino acid receptor-mediated polyphosphoionositide hydrolysis in the amygdala/pyriform cortex of amygdala-kindled rats. *Brain Res.* **485,** 95–101.

Akiyama, K., Yamada, N., and Sato, M. (1987) Increase in ibotenate-stimulated phosphatidylinositol hydrolysis in slices of the amygdala/pyriform cortex and hippocampus of rat by amygdala kindling. *Exp. Neurol.* **98,** 499–508.

Akiyama, K., Daigen, A., Yamada, N., Itoh, T., Kohira, I., Ujike, H., and Otsuki, S. (1992) Long-lasting enhancement of metabotropic excitatory amino acid receptor-mediated polyphosphoinositide hydrolysis in the amygdala/pyriform cortex of deep pyriform cortical kindled rats. *Brain Res.* 569, 72–77.

Aniksztejn, L., Otani, S., and Ben-Ari, Y. (1992) Quisqualate metabotropic receptors modulate NMDA currents and facilitate induction of long-term potentiation through protein kinase C. *Eur. J. Neurosci.* **4,** 500–505.

Artola, A. and Singer, W. (1987) Long-term potentiation and NMDA receptors in rat visual cortex. *Nature* **330,** 649–652.

Bashir, Z. I., Bortolotto, Z. A., Davies, C. H., Berretta, N., Irving, A. J., Seal, A. J., Henley, J. M., Jane, D. E., Watkins, J. C., and Collingridge, G. L. (1993a) Induction of LTP in the hippocampus needs synaptic activation of glutamate metabotropic receptors. *Nature* **363,** 347–350.

Bashir, Z. I., Jane, D. E., Sunter, D. C., Watkins, J. C., and Collingridge, G. L. (1993b) Metabotropic glutamate receptors contribute to the induction of long-term depression in the CA1 region of the hippocampus. *Eur. J. Pharmacol.* **239,** 265,266.

Baskys, A. and Malenka, R. C. (1991) *Trans*-ACPD depresses synaptic transmission in the hippocampus. *Eur. J. Pharmacol.* **193,** 131,132.

Behnisch, T. and Reymann, K. G. (1993) Co-activation of metabotropic glutamate and *N*-methyl-D-aspartate receptors is involved in mechanisms of long-term potentiation maintenance in rat hippocampal CA1 neurons. *Neuroscience* **54,** 37–47.

Bliss, T. V. P. and Lomo, T. (1973) Long-lasting potentiation of synaptic transmission in the dentate area of the anesthetized rabbits following stimulation of the perforant path. *J. Physiol. (Lond.)* **232,** 331–356.

Bortolotto, Z. A. and Collingridge, G. L. (1992) Activation of glutamate metabotropic receptors induces long-term potentiation. *Eur. J. Pharmacol.* **214,** 297,298.

Bortolotto, Z. A. and Collingridge, G. L. (1993) Characterisation of LTP induced by the activation of glutamate metabotropic receptors in area CA1 of the hippocampus. *Neuropharmacology* **32,** 1–9.

Boss, V. and Conn, P. J. (1992) Metabotropic excitatory amino acid receptor activation stimulates phospholipase D in hippocampal slices. *J. Neurochem.* **59**, 2340–2343.

Cain, D. P. (1989) Long-term potentiation and kindling: how similar are the mechanisms. *Trends Neurosci.* **12**, 6–10.

Calabresi, P., Maj, R., Pisani, A., Mercuri, N. B., and Bernardi, G. (1992) Long-term synaptic depression in the striatum: Physiological and pharmacological characterization. *J. Neurosci.* **12**, 4224–4233.

Collingridge, G. L., Kehl, S. J., and McLennan, H. (1983) Excitatory amino acids in synaptic transmission in the Schaffer collateral-commissural pathway of the rat hippocampus. *J. Physiol. (Lond.)* **334**, 33–46.

Collins, D. R. and Davies, S. N. (1993) Co-administration of (1S,3R)-1-amino-cyclopentane-1,3-dicarboxylic acid and arachidonic acid potentiates synaptic transmission in rat hippocampal slices. *Eur. J. Pharmacol.* **240**, 325,326.

Daniel, H., Hemart, N., Jaillard, D., and Crepel, F. (1992) Coactivation of metabotropic glutamate receptors and of voltage-gated calcium channels induces long-term depression in cerebellar Purkinje cells in vitro. *Exp. Brain Res.* **90**, 327–331.

Desai, M. A. and Conn, P. J. (1991) Excitatory effects of ACPD receptor activation in the hippocampus are mediated by direct effects on pyramidal cells and blockade of synaptic inhibition. *J. Neurophysiol.* **66**, 40–52.

Douglas, R. M. and Goddard, G. V. (1975) Long-term potentiation of the perforant path-granule cell synapse in the rat hippocampus. *Brain Res.* **86**, 205–215.

Frenguelli, B. G., Potier, B., Slater, N. T., Alford, S., and Collingridge, G. L. (1994) Metabotropic glutamate receptors and calcium signalling in dentrites of hippocampal CA1 neurons. *Neuropharmacol.* **32**, 1229–1237.

Gallagher, J. P. and Zheng, F. (1993) Distinct subtypes of pertussis toxin-resistant metabotropic glutamate receptors at rat dorsolateral septal nucleus neurons. *Soc. Neurosci. Abstr.* **19**, 470.

Gean, P. W., Shinnick-Gallagher, P., and Anderson, A. C. (1989) Spontaneous epileptiform activity and alteration of GABA- and NMDA-mediated neurotransmission in amygdala neurons kindled *in vivo*. *Brain Res.* **494**, 171–181.

Goddard, G. V., McIntyre, D. C., and Leech, C. K. (1969) A permanent change in brain function resulting from daily electrical stimulation. *Exp. Neurol.* **25**, 295–330.

Goddard, G. V., McNaughton, B. L., Douglas, R. M., and Barnes, C. A. (1978) Synaptic change in the limbic system; evidence from studies using electrical stimulation with and without seizure activity, in *The Continuing Evolution of the Limbic System Concept* (Livingston, K. and Hornykiewicz, O., eds.), Plenum, New York, pp. 355–368.

Goh, J. W. and Pennefather, P. S. (1989) A pertussis toxin-sensitive G protein in hippocampal long- term potentiation. *Science* **244**, 980–983.

Grover, L. M. and Teyler, T. J. (1990) Two components of long-term potentiation induced by different patterns of afferent activation. *Nature* **347**, 477–479.

Harris, E. W. and Cotman, C. W. (1986) Long-term potentiation of guinea pig mossy fiber responses is not blocked by *N*-methyl D-aspartate antagonists. *Neurosci. Lett.* **70**, 132–137.

Harris, E. W., Ganong, A. H., and Cotman, C. W. (1984) Long-term potentiation in the hippocampus involves activation of N-methyl-D-aspartate receptors. *Brain Res.* **323,** 132–137.

Harvey, J. and Collingridge, G. L. (1993) Signal transduction pathways involved in the acute potentiation of NMDA responses by 1S,3R-ACPD in rat hippocampal slices. *Br. J. Pharmacol.* **109,** 1085–1090.

Iadorola, M. J., Nicoletti, F., Naranjo, J. R., Putnam, F., and Costa, E. (1986) Kindling enhances the stimulation of inositol. *Brain Res.* **374,** 174–178.

Ito, I. and Sugiyama, H. (1991) Roles of glutamate receptors in long-term potentiation at hippocampal mossy fiber synapses. *Neuroreport* **2,** 333–336.

Ito, I., Okada, D., and Sugiyama, H. (1988) Pertussis toxin suppresses long-term potentiation of hippocampal mossy fiber synapses. *Neurosci. Lett.* **90,** 181–185.

Izumi, Y., Clifford, D. B., and Zorumski, C. F. (1991) 2-Amino-3-phosphono-propionate blocks the induction and maintenance of long-term potentiation in rat hippocampal slices. *Neurosci. Lett.* **122,** 187–190.

Johnston, D., Williams, S., Jaffe, D., and Gray, R. (1992) NMDA-receptor-independent long-term potentiation. *Ann. Rev. Physiol.* **54,** 489–505.

Kano, M. and Kato, M. (1987) Quisqualate receptors are specifically involved in cerebullar synaptic plasticity. *Nature* **325,** 276–279.

Kato, N. (1993) Dependence of long-term depression on postsynaptic metabotropic glutamate receptors in visual cortex. *Proc. Natl. Acad. Sci. USA* **90,** 3650–3654.

Katsuki, H., Kaneko, S., and Satoh, M. (1992) Involvement of postsynaptic G proteins in hippocampal long-term potentiation. *Brain Res.* **581,** 108–114.

Linden, D. J., Dickinson, M. H., Smeyne, M., and Connor, J. A. (1991) A long-term depression of AMPA currents in cultured cerebellar Purkinje neurons. *Neuron* **7,** 81–89.

Lynch, G., Larson, J., Kelso, S., Barrionuevo, G., and Schottler, F. (1983) Intracellular injection of EGTA block induction of hippocampal long-term potentiation. *Nature* **305,** 719–721.

Masu, M., Tanabe, Y., Tsuchida, K., Shigemoto, R., and Nakanishi, S. (1991) Sequence and expression of a metabotropic glutamate receptor. *Nature* **349,** 760–765.

McGuinness, N., Anwyl, R., and Rowan, M. (1991a) The effects of trans-ACPD on long-term potentiation in the rat hippocampal slice. *Neuroreport* **2,** 688–690.

McGuinness, N., Anwyl, R., and Rowan, M. (1991b) *Trans*-ACPD enhances long-term potentiation in the hippocampus. *Eur. J. Pharmacol.* **197,** 231,232.

McNamara, J. O., Rigsbee, L. C., Butler, L. S., and Shin, C. (1989) Intravenous phenytoin is an effective anticonvulsant in the kindling model. *Ann. Neurol.* **26,** 676–678.

Morris, R. G. M., Anderson, E., Lynch, G. S., and Baudry, M. (1986) Selective impairment of learning and blockade of long-term potentiation by an N-methyl-D-aspartate receptor antagonist, AP5. *Nature* **319,** 774–776.

Murphy, S. N. and Miller, R. J. (1988) A glutamate receptor regulates Ca^{2+} mobilization in hippocampal neurons. *Proc. Natl. Acad. Sci. USA* **85,** 8737–8741.

O'Connor, J. J., Rowan, M. J., and Anwyl, R. (1993a) Long-lasting potentiation of the NMDA receptor-mediated EPSC is inhibited by metabotropic glutamate receptor antagonists in rat dentate granule cells *in vitro. J. Physiol. (Lond.)* **473**, 170P.

O'Connor, J. J., Rowan, M. J., and Anwyl, R. (1993b) Potentiation of NMDA receptor-mediated excitatory postsynaptic currents following metabotropic glutamate receptor activation and tetanic stimulation in rat hippocampus. *J. Physiol. (Lond.)* **473**, 47P.

Ohishi, H., Shigemoto, R., Nakamishi, S., and Mizuno, N. (1993) Distribution of the messenger RNA for a metabotropic glutamate receptor, mGluR2, in the central nervous system. *Neuroscience* **53**, 1009–1018.

Ohishi, H., Shigemoto, R., Nakamishi, S., and Mizuno, N. (1993) Distribution of the mRNA for a metabotropic glutamate receptor (mGluR3) in the rat brain. *J. Comp. Neurol.* **335**, 252–266.

Otani, S. and Ben-Ari, Y. (1991) Metabotropic receptor-mediated long-term potentiation in rat hippocampal slices. *Eur. J. Pharmacol.* **205**, 325,326.

Otani, S., Ben-Ari, Y., and Roisin-Lallemand, M.-P. (1993) Metabotropic receptor stimulation coupled to weak tetanus leads to long-term potentiation and a rapid elevation of cytosolic protein kinase C activity. *Brain Res.* **613**, 1–9.

Racine, R. (1972) Modification of seizure activity by electrical stimulation: I. Afterdischarge threshold. *Electroencephalogr. Clin. Neurophysiol* **32**, 269–279.

Radpour, S. and Thomson, A. M. (1992) Synaptic enhancement induced by NMDA and Q_p receptors and presynaptic activity. *Neurosci. Lett.* **138**, 119–122.

Rainnie, D. G. and Shinnick-Gallagher, P. (1992) Activation of postsynaptic metabotropic receptors by trans-ACPD evokes a membrane hyperpolarization in neurons of the basolateral amygdala. *Soc. Neurosci. Abstr.* **22**, 266.

Rainnie, D. G., Asprodini, E., and Shinnick-Gallagher, P. (1992) Kindling-induced long-lasting changes in excitatory and inhibitory transmission in the basolateral amygdala. *J. Neurophysiol.* **67**, 443–454.

Rainnie, D. G., Holmes, K. H. and Shinnick-Gallagher, P. (1994) Activation of postsynaptic metabotropic glutamate receptors by trans-ACPD hyperpolarizes neurones of the basolateral amygdala. *J. Neurosci.*, in press.

Reymann, K. G. and Matthies, H. (1989) 2-Amino-4-phosphonobutyrate selectively eliminates late phases of long-term potentiation in rat hippocampus. *Neurosci. Lett.* **98**, 166–171.

Riedel, G. and Reymann, K. (1993) An antagonist of the metabotropic glutamate receptor prevents LTP in the dentate gyrus of freely moving rats. *Neuropharmacology* **32**, 929–931.

Sacaan, A. I. and Schoepp, D. D. (1992) Activation of hippocampal metabotropic excitatory amino acid receptors leads to seizures and neuronal damage. *Neurosci. Lett.* **139**, 77–82.

Sato, M., Racine, R. J., and McIntyre, D. C. (1990) Kindling: basic mechanisms and clinical validity. *Electroencephalogr. Clin. Neurophysiol* **76**, 459–472.

Savage, D. D., Werling, L. L., Nadler, J. V., and McNamara, J. O. (1984) Selective and reversible increase in the number of quisqualate-sensitive glutamate binding sites on hippocampal synaptic membranes after angular bundle kindling. *Brain Res.* **307**, 332–335.

Schmidt, J. T. (1990) Long-term potentiation and activity-dependent retinotopic sharpening in the regenerating retinotectal projection of goldfish: common sensitive period and sensitivity to NMDA blockers. *J. Neurosci.* **10**, 233–246.

Schoepp, D. D. and Johnson, B. G. (1988) Excitatory amino acid agonist-antagonist interactions at 2-amino-4-phosphonobutyric acid-sensitive quisqualate receptors coupled to phosphoinositide hydrolysis in slices of rat hippocampus. *J. Neurochem.* **50**, 1605–1613.

Schoepp, D. D. and Johnson, B. G. (1989) Comparison of excitatory amino acid-stimulated phosphoinositide hydrolysis and N-'3H:acetylaspartylglutamate binding in rat brain: selective inhibition of phosphoinositide hydrolysis by 2-amino-3-phosphonopropionate. *J. Neurochem.* **53**, 273–278.

Schoepp, D. D., Johnson, B. G., and Monn, J. A. (1992) Inhibition of cyclic AMP formation by a selective metabotropic glutamate receptor agonist. *J. Neurochem.* **58**, 1184–1186.

Schoepp, D. D., Johnson, B. G., True, R. A., and Monn, J. A. (1991) Comparison of (1S,3R)-1-aminocyclopentane-1,3-dicarboxylic acid (1S,3R-ACPD)- and 1R,3S-ACPD-stimulated brain phosphoinositide hydrolysis. *Eur. J. Pharmacol. Mol. Pharmacol.* **207**, 351–353.

Shigemoto, R., Nakanishi, S., and Mizuno, N. (1992) Distribution of the mRNA for a metabotropic glutamate receptor (mGluR1) in the central nervous system: An *in situ* hybridization study in adult and developing rat. *J. Comp. Neurol.* **322**, 121–135.

Sladeczek, F., Pin, J. P., Recasens, M., Bockaert, J., and Weiss, S. (1985) Glutamate stimulates inositol phosphate formation in striatal neurons. *Nature* **317**, 717–719.

Stanton, P. K., Chattarji, S., and Sejnowski, T. J. (1991) 2-Amino-3-phosphonopropionic acid, an inhibitor of glutamate-stimulated phosphoinositide turnover, blocks induction of homosynaptic long-term depression, but not potentiation, in rat hippocampus. *Neurosci. Lett.* **127**, 61–66.

Tanabe, Y., Masu, M., Ishii, T., Shigemoto, R., and Nakanishi, S. (1992) A family of metabotropic glutamate receptors. *Neuron* **8**, 169–179.

Tanabe, Y., Nomura, A., Masu, M., Shigemoto, R., Mizuno, N., and Nakanishi, S. (1993) Signal transduction, pharmacological properties, and expression patterns of two rat metabotropic glutamate receptors, mGluR3 and mGluR4. *J. Neurosci.* **13**, 1372–1378.

Tsubokawa, H., Robinson, H. P. C., Takenawa, T., and Kawai, N. (1993) Block of long-term potentiation by intracellular application of anti-phosphatidylinositol 4,5-bisphosphate antibody in hippocampal pyramidal neurons. *Neuroscience* **55**, 643–651.

Vecil, G. G., Li, P. P., and Warsh, J. J. (1992) Evidence for metabotropic excitatory amino acid receptor heterogeneity: Developmental and brain regional studies. *J. Neurochem.* **59**, 252–258.

Williams, S. and Johnston, D. (1988) Muscarinic depression of long-term potentiation in CA3 hippocampal neurons. *Science* **242**, 84–87.

Yamada, N., Akiyama, K., and Otsuki, S. (1989) Hippocampal kindling enhances excitatory amino acid receptor-mediated polyphosphoinositide hydrolysis in the hippocampus and amygdala/pyriform cortex. *Brain Res.* **490**, 126–132.

Zalutsky, R. A. and Nicoll, R. A. (1990) Comparison of two forms of long-term potentiation in single hippocampal neurons. *Science* **248,** 1619–1624.

Zalutsky, R. A. and Nicoll, R. A. (1992) Mossy fiber long-term potentiation shows specificity but no apparent cooperativity. *Neurosci. Lett.* **138,** 193–197.

Zheng, F. and Gallagher, J. P. (1990) Long-term potentiation (LTP) in rat dorsal lateral septal nucleus (DLSN) is *not* blocked by D,L-2-amino-5-phosphono-pentanoate (AP5). *Soc. Neurosci. Abstr.* **16,** 653.

Zheng, F. and Gallagher, J. P. (1992a) Calcium release from internal stores is required for the induction of metabotropic glutamate receptor-dependent long-term potentiation in dorsolateral septal nucleus neurons *in vitro. Soc. Neurosci. Abstr.* **18,** 642.

Zheng, F. and Gallagher, J. P. (1992b) Metabotropic glutamate receptors are required for the induction of long-term potentiation. *Neuron* **9,** 163–172.

Zheng, F. and Gallagher, J. P. (1993) Pertussis toxin selectively blocked burst firing induced by 1S,3R-ACPD at rat dorsolateral septal nucleus neurons. *Soc. Neurosci. Abstr.* **19,** 470.

Zheng, F., Lonart, G., Johnson, K. M., and Gallagher, J. P. (1994) (1S,3R)-1-amino-cyclopentane-1,3-dicarboxylic acid (1S,3R-ACPD) induced burst firing via an inositol-1,4,5-triphosphate-independent pathway at rat dorsolateral septal nucleus. *Neuropharmacology* **33,** 97–102.

CHAPTER 8

Regulation of Neuronal Circuits and Animal Behavior by Metabotropic Glutamate Receptors

P. Jeffrey Conn, Danny G. Winder, and Robert W. Gereau IV

1. Introduction

Within the nervous system, neurons are organized into circuits or networks of interconnected cells that mediate specific functional tasks. Traditionally, most of the major connections that make up a neural network have been viewed as synaptic connections in which activation of ligand-gated ion channels results in either excitation or inhibition of a receptive neuron (i.e., fast synaptic transmission). However, in recent years, the increase in our understanding of second messengers and their roles in modulating neuronal excitability and synaptic transmission has led to an enlargement of this concept. It is now clear that fast synaptic transmission through networks of neurons can be modulated by activation of receptors coupled to second-messenger systems through GTP-binding proteins. For instance, in

The Metabotropic Glutamate Receptors Eds.: P. J. Conn and J. Patel
© 1994 Humana Press Inc., Totowa, NJ

a network of neurons connected by glutamatergic synapses, it was generally held that glutamate would elicit fast synaptic responses by activating members of the ionotropic glutamate receptor family. Neuromodulators from extrinsic afferents (i.e., acetylcholine, serotonin, norepinephrine, and so forth) could then modulate transmission through the network of glutamatergic neurons by activating GTP-binding protein-linked receptors and second-messenger systems. Although there are a number of neurotransmitters that activate both ligand-gated ion channels and receptors coupled to second-messenger systems, until recently, it was thought that all of the actions of glutamate, the major excitatory neurotransmitter in the brain, were mediated by activation of ionotropic glutamate receptors and generation of fast synaptic responses. Because glutamate is the major excitatory neurotransmitter in the brain, separation of synapses involved in generation of fast synaptic responses and modulatory control of transmission through a neural circuit was thought to be the rule in the central nervous system, with relatively few examples of synapses in which a neurotransmitter elicits both fast and slow synaptic responses at a single synapse. The discovery of metabotropic glutamate receptors (mGluRs) dramatically alters the traditional view of regulation of transmission in the central nervous system. As discussed in detail in previous chapters, activation of mGluRs exerts a number of effects on neuronal excitability and synaptic transmission previously associated only with neuromodulators from nonglutamatergic afferents (*see* Chapters 5 and 6). Thus, glutamate may activate G-protein-linked receptors at the same synapses at which it elicits fast synaptic responses and thereby set the gain at the synapse. Furthermore, as discussed in Chapter 10, the sensitivity of mGluRs is developmentally regulated and is altered under certain pathological conditions, including ischemia and epilepsy. Such regulation of metabotropic glutamate receptor function may dramatically influence net transmission through networks involving glutamatergic synapses under physiological and pathological conditions.

    At present, our understanding of the roles of mGluRs in modulating neuronal function is largely restricted to changes at the cellular level, with a less detailed understanding of the net effect of these cellular changes on transmission through intact neuronal circuits and animal behavior. Ultimately, the precise roles of specific mGluR subtypes in regulating transmission through circuits containing glutamatergic synapses will require the development of specific mGluR antagonists, so that the influence of specific mGluR subtypes on transmission through the circuit can be removed. However, studies of the physiological effects of the selective mGluR agonist *trans*-ACPD and its 1S,3R-isomer have resulted in a great deal of insight into the roles of mGluRs regulating transmission through intact neuronal circuits. Rather than attempt to give an exhaustive review of the physiological roles of mGluRs in various brain regions, the intent of this chapter is to focus on specific neuronal circuits in the brain where the roles of mGluRs have been especially well characterized or where the behavioral relevance of mGluR activation has been directly studied. These include the hippocampus, striatum, nucleus of the solitary tract, cerebellum, and retina.

2. Regulation of Transmission Through the Hippocampal Trisynaptic Circuit

The hippocampus is a cortical structure that is thought to play an important role in processing of complex spacial and temporal patterns, and formation of short- and long-term memory. In addition, the hippocampus is a primary target of certain neurological disorders, such as Alzheimer's disease and temporal lobe epilepsy. Because of this, much effort has been directed at developing a detailed understanding of the synaptic organization of the hippocampus as well as the cellular mechanisms involved in regulation of synaptic transmission in this structure (Brown and Zador, 1990). The hippocampus is commonly viewed as a relatively simple circuit consisting of three major excitatory synapses (Fig. 1). The primary input to the hippocampus is from the entorhinal cortex, which sends excitatory afferents to the dentate gyrus (DG) via the perforant path. Mossy fibers from the dentate granule cells project to area CA3 of the hippocampus proper, which in turn sends afferents via the Schaffer collateral to hippocampal area CA1. Afferents from CA1 pyramidal cells then

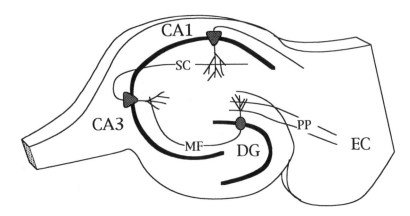

Fig. 1. Schematic illustration of a transverse section of the hippocampal formation showing the major components of the hippocampal trisynaptic circuit. EC = entorhinal cortex, PP = perforant path, DG = dentate gyrus, MF = mossy fibers, SC = Schaffer collateral.

provide the major output of the hippocampus. Glutamate is likely to be the neurotransmitter at each of the three major excitatory synapses in the hippocampal formation. In addition to this overall trisynaptic circuitry, each major subdivisions of the hippocampal formation (CA1, CA3, and dentate gyrus) has its own intrinsic circuitry that includes the principal neurons of the region as well as different populations of inhibitory interneurons that are responsible for feedback and feedforward inhibition. This is an oversimplified view of hippocampal circuitry, but provides a useful framework for discussion of regulation of transmission through the hippocampal formation.

2.1. Multiple mGluR Subtypes Are Heterogeneously Distributed Throughout the Hippocampal Formation

The earliest studies of mGluR function in mammalian CNS demonstrated that mGluR agonists elicit a large phosphoinositide hydrolysis response in hippocampal slices (Nicolletti et al., 1986) and that this response is more robust in the hippocampus than in other brain regions (Desai and Conn, 1990). Since that time, *in situ* hybridization and immunocytochemical studies as well as biochemical and electrophysiological studies have revealed that multiple mGluR subtypes are present in the hippocampus and that these receptors are heterogeneously distributed in different subregions of the hippocampal formation (*see* Chapter 4 for detailed discussion). mGluR5 and mGluR7 are the most abundantly expressed of the cloned mGluRs in the hippocampus, and are highly expressed in most subregions of the hippocampal formation (Abe et al., 1992; Fotuhi et al., 1993; Okamoto et al., 1994). mGluR1 is expressed predominantly in the dentate gyrus and area CA3, but is also present in high levels in neurons at the border of the stratum oriens and alveus of area CA1 (Houamed et al., 1991; Masu et al., 1991; Martin et al., 1992; Shigemoto et al., 1992; Fotuhi et al., 1993). Preliminary immunocytochemical studies suggest that mGluR1a is localized in postsynaptic membranes (Martin et al., 1992). mGluR2 and mGluR3 are predominantly expressed in the dentate gyrus (Tanabe et al., 1992; 1993; Fotuhi et el., 1993;). mRNA for mGluR4 is virtually absent from all hippocampal subregions, though low levels of mGluR4 expression can be detected in area CA2 (Tanabe et al., 1993; Fotuhi et el., 1993; Kristensen et al., 1993).

Although it is not yet clear what effector systems are employed by specific mGluR subtypes in the hippocampus, biochemical studies suggest that mGluRs in the hippocampal formation are coupled to a variety of second-messenger systems, including activation of phosphoinositide hydrolysis, increased cAMP accumulation, and inhibition of adenylyl cyclase. In addition, a novel metabotropic EAA receptor that is selectively activated by L-cysteine sulfinic acid has recently been characterized that is coupled to activation of phospholipase D (*see* Boss et al., 1994; Chapter 3).

2.2. Effects of mGluR Activation on Transmission Through the Hippocampal Circuit

2.2.1. mGluR Activation Reduces Excitatory Synaptic Responses at the Perforant Path–Dentate Gyrus Synapse

Concentrations of *trans*-ACPD or its active isomer (1S,3R-ACPD) that selectively activate mGluRs induce pronounced effects on excitatory synaptic responses at each of the major synapses of the hippocampal trisynaptic circuit. Extracellular field potential recordings reveal that *trans*-ACPD decreases evoked population spikes at the perforant path–dentate gyrus synapse and, thus, reduces the net input into the hippocampal formation from the entorhinal cortex (Desai and Conn, 1991). Since population spikes represent the simultaneous firing of a population of neurons in the immediate vicinity of the recording electrode, this indicates that fewer dentate granule cells fire in response to perforant path stimulation when mGluRs are activated than when they are inactive. Although the mechanism of this effect has not been examined in detail, the decrease in evoked population spikes is accompanied by a reduction of excitatory postsynaptic currents (EPSCs) (Baskys et al., 1993; Winder and Conn, unpublished findings), and is likely to be mediated by activation of a presynaptic autoreceptor and reduction of glutamate release (*see* Chapter 6). Because of this action, a weak signal that arrives at the dentate gyrus is less likely to be transmitted to areas CA3 and CA1.

2.2.2. mGluR Activation Increases Excitability of CA3 Pyramidal Cells

In contrast to its effects in the dentate gyrus, mGluR activation increases the amplitude of population spikes at the mossy fiber–CA3 synapse (Desai and Conn, 1991). This increase in population spike amplitude is at least partially mediated by a variety of direct excitatory effects of mGluR activation on CA3 pyramidal cells (Charpak et al., 1990) (*see* Chapter 5). These include cell depolarization (likely owing to blockade of a leak potassium current), blockade of spike frequency adaptation (owing to blockade of calcium-dependent potassium current termed I_{AHP}), and blockade of the slow voltage-dependent potassium current I_M. The effects of mGluR activation on CA3 pyramidal cell excitability can be mimicked by high-frequency stimulation of the mossy fiber pathway (Charpak and Gahwiler, 1991). This finding is consistent with the hypothesis that mGluRs involved in this response can be synaptically activated. However, at the time of these studies, mGluR antagonists were not available, and the area of the hippocampal slice that was stimulated also contained afferents from nonglutamatergic neurons. Thus, it is not entirely clear whether the effects of mossy fiber stimulation are mediated by activation of mGluRs. In addition to the direct excitatory effects on CA3 pyramidal cells, evidence suggests that activation of mGluRs can lead to induction of long-term potentiation (LTP) of synaptic transmission at the mossy fiber–CA3 synapse (Bashir et al., 1993). Thus, most of the actions of mGluR activation in area CA3 may ultimately serve to increase transmission at this point in the hippocampal trisynaptic circuit. However, mGluR activation also causes direct excitation of inhibitory interneurons in area CA3, and this effect can be mimicked by stimulation of mossy fiber afferents (Miles and Poncer, 1993). It is not yet known how this affects the overall actions of mGluR activation in area CA3.

2.2.3. mGluR Agonists Have Multiple Actions in Hippocampal Area CA1

The effects of mGluR activation in area CA1 are more complex than in the dentate gyrus or area CA3 in that mGluR agonists have effects that can both increase and decrease net transmission through this point in the hippocampal trisynaptic circuit. mGluR activation reduces transmission at the Schaffer collateral–CA1 synapse, probably by activation of a presynaptic autoreceptor (Baskys and Malenka, 1991a,b; Desai et al., 1992). However, mGluR agonists also have direct excitatory effects on CA1 pyramidal cells that are similar to their effects on pyramidal cells in area CA3 (Stratton et al., 1989; Desai and Conn, 1991; Hu and Storm, 1991, 1992). In addition, mGluR activation reduces synaptic inhibition in area CA1. This can be seen as a reduction of both $GABA_A$ and $GABA_B$ receptor-mediated IPSPs recorded in CA1 pyramidal cells after stimulation of the Schaffer collateral (Pacelli and Kelso, 1991; Desai and Conn, 1991, 1992). Detailed analysis of the effects of mGluR agonists on different components of the circuitry involved in generation of IPSPs in area CA1 suggests that this effect is mediated by a reduction of transmission at both inhibitory synapses onto hippocampal pyramidal cells and excitatory synapses onto inhibitory interneurons (Desai and Conn, 1992). A number of other neuromodulators have actions similar to those of mGluR agonists in that they reduce synaptic inhibition in area CA1 and thereby induce a pronounced increase in net excitatory drive through the Schaffer collateral–CA1 synapse (Madison and Nicoll, 1988; Doze et al., 1991; Oleskevich and Lacaille, 1992).

As in area CA3, activation of mGluRs in area CA1 can induce LTP or potentiate induction of LTP by a tetanic stimulus (see Chapter 7). This is likely to be at least partially mediated by an mGluR-induced potentiation of conductance through the N-methyl-D-aspartate (NMDA) subtype of glutamate receptor (Aniksztejn et al., 1992). Also, activation of mGluRs in the hippocampus potentiates β-adrenergic receptor-mediated increases in cAMP accumulation (Winder and Conn, 1993; Winder et al., 1993) and coactivation of mGluRs and β-adrenergic receptors in area CA1 leads to a long-lasting (>30 min) increase in the excitability of CA1 pyramidal cells that is mediated by cAMP (Gereau and Conn, 1994; see Chapter 3). Finally, although mGluR-mediated disinhibition is transient in area CA1 in slices from rats (Desai and Conn, 1991; Pacelli and Kelso, 1991), mGluR activation leads to a long-lasting reduction in synaptic inhibition in area CA1 from rabbit (Liu et al, 1993). Thus, in addition to its acute physiological effects in area CA1, activation of mGluRs has a number of effects that lead to a long-lasting increase in transmission through this portion of the hippocampal circuit.

2.2.4. The Combined Actions of mGluR Activation
May Increase Signal-to-Noise Ratio
of Transmission Through the Hippocampal Formation

The studies discussed above suggest that mGluR activation has opposing effects on transmission through the hippocampal trisynaptic circuit. Acutely, mGluR agonists reduce excitatory transmission in the dentate gyrus and area CA1 of the hippocampus, but simultaneously increase excitability of hippocampal pyramidal neurons (by both direct excitatory effects and disinhibition). The net effect of these combined inhibitory and excitatory effects is not entirely clear. However, one possibility is that these combined actions serve to increase the signal-to-noise ratio of transmission through the hippocampus. Because of actions of mGluR activation in the dentate gyrus, weak stimuli that reach the dentate gyrus under conditions in which mGluRs are activated would be filtered out and would be less likely to be transmitted to area CA3. However, if a stimulus reaches the dentate gyrus that is strong enough to be transmitted to area CA3, the increase in excitability of pyramidal cells in area CA3 would result in the signal being amplified in this region, and a stronger signal would then be transmitted to area CA1. In area CA1, the multiple actions of mGluR activation could further increase the signal-to-noise ratio. Because of the mGluR-mediated reduction in transmission at the Schaffer collateral–CA1 synapse, weak stimuli that arrive at this synapse would once again be filtered out and would be less likely to elicit action potentials in CA1 pyramidal cells. However, if a strong, suprathreshold stimulus reaches area CA1, the increased pyramidal cell excitability and reduced synaptic inhibition brought about by mGluR activation would amplify the response of these cells and thereby increase the net output of CA1 pyramidal cells. Such modulation of signal-to-noise ratio has been shown to occur with some other neurotransmitters and has been proposed to play an important role in regulation of attentiveness to sensory stimuli (Madison and Nicoll, 1986). The hippocampus is known to play an important role in processing of complex spacial and temporal patterns, and individual hippocampal cells respond specifically to individual words, particular faces, or particular spacial coordinates of their environment (Brown and Zador, 1990). It is possible that, by screening out some stimuli while allowing or enhancing responses to others, mGluRs may play an important role in allowing hippocampal neurons to respond selectively to particular environmental cues. However, it should be noted that the studies discussed above involve measurement of responses to simultaneous activation of multiple mGluRs by bath application of nonselective mGluR agonists. It is not yet clear whether conditions exist in vivo where multiple mGluRs will be simultaneously activated. Tonic activation of mGluRs by extracellular glutamate (Sah et al., 1989) may provide a mechanism by which mGluRs involved in each of the actions discussed above could be activated simultaneously. Alternatively, there may be a difference in the spatial and temporal sequence of mGluR activation at different sites in the hippocampal formation.

2.2.5. Activation of mGluRs in the Hippocampus May Contribute to Induction of Seizure Activity

In addition to playing a role in normal physiological processes, such as learning and memory, the hippocampus is a highly seizure-prone structure that is commonly involved in temporal lobe epilepsy (Lothman et al., 1991). Because of this, much effort has been directed at determining the cellular mechanisms of synchronization of hippocampal neuronal activity and spread of seizure activity through this structure. Previous reports suggest that development of synchronized bursting in the hippocampus is associated with a reduction in synaptic inhibition and that this is important for transition to the ictal state. Although it is likely that multiple factors contribute to the loss of inhibition in hippocampal area CA1 that occurs with generation of seizure activity, mGluR-mediated disinhibition could play a significant role. Furthermore, the direct excitatory effects of mGluR activation on hippocampal pyramidal cells and potentiation of NMDA receptor currents (see Chapters 5 and 6) could contribute to generation of synchronized bursting. Consistent with this, extracellular field potential recordings reveal that trans-ACPD results in generation of multiple population spikes in area CA1 of the hippocampus after stimulation of the Schaffer collateral (Desai and Conn, 1991). In a normal healthy hippocampus, stimulation of the Schaffer collateral elicits a single population spike, and generation of multiple population spikes is indicative of seizure activity. Sacaan and Schoepp (1992) recently reported that microinjection of 1S,3R-ACPD into the hippocampus results in delayed-onset development of temporal lobe seizures and fully generalized behavioral convulsions in rats. Thus, it is possible that mGluR-mediated disinhibition and excitation of hippocampal pyramidal cells participate in temporal lobe epilepsy. If this is the case, selective antagonists of the mGluR subtypes involved in these seizure-promoting effects in the hippocampus could provide a novel treatment for certain types of epileptic disorders.

3. Roles of mGluRs in Regulation of Cerebellar Function

A second major region of the central nervous system in which the roles of mGluRs have been particularly well characterized is the cerebellum. The cerebellum interacts extensively with both sensory and motor systems, and plays an essential role in fine control of motor function (for reviews, *see* Llinas and Walton, 1990; Ghez, 1991). However, the cerebellum is not a major relay station for sensory or motor function, and complete ablation of this structure does not produce sensory impairment or a reduction in motor strength. Instead, lesions of the cerebellum dramatically compromise balance and coordination of limb and eye movements, and impair the ability of the organism to execute even the simplest of motor sequences. Thus, the cerebellum plays an essential role in fine control of coordinated movement. The cerebellum is likely to accomplish this task by continuously comparing commands for movement from higher centers with information about actual movements from sensory feedback received from the periphery. In addition to its role in acute regulation of motor function, long-lasting changes in synaptic transmission at specific sites in the cerebellar circuit are thought to play an important role in motor memory. A variety of disorders have been described that impair cerebellar function and have a dramatic effect on equilibrium, speech, and/or coordinated movement.

3.1. Basic Circuitry of the Cerebellum

The circuitry of the cerebellum has been reviewed in detail (Llinas and Walton, 1990; Ghez, 1991). Below is a brief outline of the major aspects of cerebellar circuitry (*see* Fig. 2). The cerebellum is composed of the cerebellar cortex (which forms an outer mantle of gray matter) and three pair of deep nuclei. Both the cerebellar cortex and the deep nuclei receive excitatory input from three major extrinsic sources: the periphery, the brainstem, and the cerebral cortex. The large Purkinje cells are the principal neurons of the cerebellar cortex and project almost exclusively to the deep nuclei of the cerebellum. The Purkinje cells are inhibitory neurons and provide a strong inhibi-

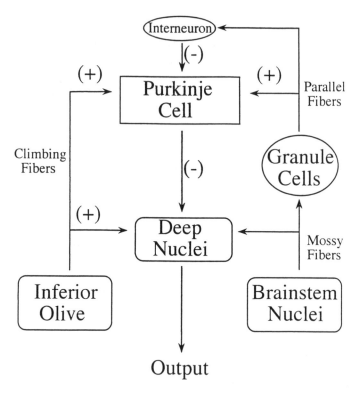

Fig. 2. Diagram showing the major components of cerebellar circuitry. Excitatory and inhibitory pathways are denoted by (+) and (−) signs, respectively.

tory input to cerebellar nuclei. The deep nuclei then provide the major output of the cerebellum, and project primarily to motor systems of the cerebral cortex and the brainstem.

Projections to the cerebellum arrive via two major excitatory pathways: the mossy fiber pathway and the climbing fiber pathway. The largest input to the cerebellum comes via the mossy fibers, which originate (directly or indirectly) from many brain regions, including brainstem nuclei, the spinal cord, and the cerebral cortex. The mossy fibers send collaterals directly to the cerebellar nuclei before synapsing on excitatory interneurons in the cerebellar cortex referred to as granule cells. The granule cells lie directly below the Purkinje cells and project to the Purkinje cells via the parallel fibers. A single parallel fiber can intersect several thousand Purkinje cell dendrites. Thus, a single mossy fiber can influence a large number of Purkinje cells.

The climbing fibers originate from a single brainstem nucleus, the inferior olivary nucleus. These fibers project into the cerebellar cortex, where each fiber branches repeatedly and wraps around the Purkinje cell soma and dendrites, giving the appearance of climbing along the Purkinje cell dendritic tree. The climbing fibers also send collaterals directly to cells within the deep cerebellar nuclei. In contrast to the mossy fiber system, each climbing fiber makes synaptic contact with only one to ten Purkinje cells. Also, each Purkinje cell receives input from only one climbing fiber. The climbing fibers provide an unusually strong excitatory input to the Purkinje cells, and a single action potential in the climbing fiber elicits a large excitatory postsynaptic potential with a high-frequency burst of action potentials referred to as the complex spike. In contrast, the parallel fibers elicit small EPSPs in Purkinje cells, and summation of multiple parallel fiber EPSPs is required to elicit a single Purkinje cell action potential.

3.2. Distribution of mGluRs in the Cerebellum

Multiple mGluR subtypes are localized in the cerebellum, and each of the major populations of cerebellar neurons expresses one or more of the cloned mGluR subtypes (*see* Chapter 4). *In situ* hybridization studies reveal that the highest density of staining for mGluR1 mRNA in rat brain is in the Purkinje cells of the cerebellar cortex. mRNA for mGluR1 is also found in more moderate levels in granule, Golgi, and stellate cells of the cerebellar cortex as well as most neurons in the deep cerebellar nuclei (Shigemoto et al., 1992). Studies with mGluR subtype-specific antibodies and specific mGluR1c probes suggest that mGluR1a (Martin et al., 1992; Fotuhi et al., 1993), mGluR1b (Fotuhi et al., 1993), and mGluR1c (Pin et al., 1992) are all expressed in the cerebellum. The highest expression of mGluR2 in rat brain is also in the cerebellum (Tanabe et al., 1992; Ohishi et al., 1993b). However, cerebellar expression of both mGluR2 and mGluR3 is largely restricted to Golgi cells with little detectable expression in other cerebellar neurons (Tanabe et al., 1992, 1993; Ohishi et al., 1993a,b). Likewise, mGluR5 expression in the cerebellum is restricted to Golgi cells, but staining for mGluR5 mRNA is only detected in a small subpopulation of these cells (Abe et al., 1992). Expression of mGluR4 in the cerebellum is largely restricted to granule cells (Kristensen et al., 1993; Tanabe et al., 1993).

Evidence suggests that mGluRs exist in the cerebellum that are coupled to increases in phosphoinositide hydrolysis (Blackstone et al., 1989; Littman et al., 1993) and increases in cGMP accumulation (Okada, 1992). Also, although mGluR agonists do not inhibit forskolin-stimulated cAMP accumulation in cerebellar slices (Genazzani et al., 1993), mGluR activation does reduce forskolin-stimulated cAMP accumulation in cultured cerebellar granule cells (Wroblewska et al., 1993). Studies in cerebellar slices suggest that the cerebellum does not contain mGluRs coupled to potentiation of cAMP responses to other agonists (Winder et al., 1993) or metabotropic L-CSA receptors coupled to activation of phospholipase D (Boss and Conn, 1992,; Boss et al., 1994, unpublished findings).

3.3. Physiological Roles of mGluRs in the Cerebellar Cortex

3.3.1. mGluRs May Participate in Transmission at the Parallel Fiber-Purkinje Cell Synapse

A large number of biochemical and electrophysiological studies implicate glutamate as the neurotransmitter at the parallel fiber–Purkinje cell synapse (*see* Llinas and Walton, 1990 for review). The fast excitatory synaptic response at this synapse is apparently mediated by the non-NMDA subtypes of ionotropic glutamate receptors and can be blocked by application of selective ionotropic glutamate receptor antagonists. Batchelor and Garthwaite (1993) recently investigated the possibility that mGluRs also participate in transmission at this synapse by measuring evoked synaptic potentials in Purkinje cells in the presence of ionotropic glutamate receptor antagonists. Under these conditions, stimulation of parallel fibers gave rise to two slow potentials. The first was a depolarization that peaked at about 400 ms after the start of the stimulation. This was followed by a slow hyperpolarizing potential that peaked about 30 s after stimulation. A series of studies suggested that these potentials were synaptically mediated. Furthermore, the sequence of potentials is mimicked by perfusion of the slices with 1S,3R-ACPD (Crepel et al., 1991; Glaum et al., 1992; Batchelor and Garthwaite, 1993), consistent with the hypothesis that they are mediated by activation of mGluRs. Interestingly, these late synaptic potentials could not be elicited by single stimuli, but were maximal after six shocks in a high-frequency train. Thus, mGluRs may not participate in postsynaptic responses to low-frequency activity of parallel fibers, but may be activated by a high-frequency burst of action potentials. The biphasic nature of the putative mGluR-mediated response may serve to give a prolonged period of excitation (relative to that elicited by ionotropic glutamate receptor activation alone) in response to high-frequency activity of the parallel fibers, followed by an even longer period of inhibition.

Consistent with the studies of evoked synaptic potentials, voltage-clamp studies in cerebellar slices reveal that mGluR agonists induce an inward current followed by an outward current in Purkinje cells (Staub et al., 1992; Takagi et al., 1992; Vranesic et al., 1993). The initial depolarization or inward current is accompanied by an increase in cell conductance (Crepel et al., 1991; Glaum et al., 1992; Staub et al., 1992), but does not have properties consistent with its being mediated by an increase in a Na^+, Ca^{2+}, or Cl^- conductance (Staub et al., 1992). However, the inward current is markedly reduced by substitution of extracellular Na^+ with Li^+ or choline, or by intracellular injection of Ca^{2+} chelators. Based on this, it was suggested that this current is mediated by activation of a Na^+/Ca^{2+} exchanger or a Ca^{2+}-dependent cation channel (Staub et al., 1992). In contrast, the late outward current is accompanied by a decrease in cell conductance and has properties that suggest it is mediated by inhibition of a voltage-independent inward current that is active at rest (Vranesic et al., 1993).

3.3.2. Activation of mGluRs on Granule Cells Reduces Excitatory Drive at the Parallel Fiber–Purkinje Cell Synapse

In addition their role in mediating the postsynaptic actions of glutamate on Purkinje cells, evidence suggests that activation of mGluRs has actions that acutely decrease excitatory input to the Purkinje cells through the parallel fiber pathway. Application of 1S,3R-ACPD to cerebellar slices results in hyperpolarization of cerebellar granule cells (East and Garthwaite, 1992). This is likely to be at least partially mediated by activation of phosphoinositide hydrolysis with subsequent calcium release and activation of a large conductance Ca^{2+}-dependent K^+ channel (Fagni et al., 1991). In addition, mGluR autoreceptors may exist on parallel fiber terminals that are involved in regulation of glutamate release. Application of 1S,3R-ACPD to cerebellar slices induces a pronounced reduction in the amplitude and initial slope of evoked EPSPs and excitatory postsynaptic currents (EPSCs) at the parallel fiber–Purkinje cell synapse (Crepel et al., 1991; Glaum et al., 1992). Though the mechanism of this response has not been characterized in detail, it is interesting to note that mGluR4 has been proposed to be an L-AP4-sensitive autoreceptor and is highly expressed in cerebellar granule cells, but not in other cerebellar neurons (Kristensen et al., 1993; Tanabe et al., 1993). Furthermore, ACPD increases, rather than decreases, current responses to exogenously applied AMPA (Glaum et al., 1992), consistent with the hypothesis that this action is mediated by activation of a presynaptic autoreceptor. However, if the reduction in evoked EPSCs is mediated by a decrease in glutamate release, ACPD might be expected to reduce the frequency of miniature EPSCs (mEPSCs) simultaneously. In contrast, analysis of mEPSCs in Purkinje cells reveals that ACPD induces a paradoxical increase in mEPSC frequency (Takagi et al., 1992). Although other possible explanations for these findings exist, it is possible that the spontaneous mEPSCs recorded in these cells may reflect spontaneous release at climbing fiber synapses rather than parallel fiber synapses. If so, mGluR activation may depress transmission at parallel fiber synapses while simultaneously increasing transmission at climbing fiber synapses. In future experiments, it will be important to determine the effects of mGluR activation at the climbing fiber synapse directly.

In addition to presynaptic modulation of synaptic transmission, activation of mGluRs may also modulate synaptic transmission postsynaptically in the cerebellum. As mentioned above, Glaum et al. (1992) found that activation of mGluRs resulted in increased current elicited by exogenous AMPA. Two lines of evidence suggest that the increase in AMPA receptor currents is mediated by effects on synaptic, rather than extrasynaptic AMPA receptor channels. First, the ACPD-induced reduction in EPSCs is sometimes preceded by a transient increase in EPSCs (Glaum et al., 1992). This could reflect an increase in AMPA receptor responses that precedes a more dramatic decrease in glutamate release. Second, ACPD increases the amplitude of mEPSCs in Purkinje cells (Takagi et al., 1992). Since mGluR activation is unlikely to increase quantal size, this effect is likely to be mediated by a postsynaptic increase in AMPA receptor responses.

3.3.3. mGluRs May Participate in Long-Term Depression of Transmission at the Parallel Fiber–Purkinje Cell Synapse

An interesting property of the parallel fiber–Purkinje cell synapse is that transmission at this synapse can undergo a long-term depression (LTD) if the Purkinje cells are simultaneously activated by parallel fibers and climbing fibers (*see* Ito, 1989 for review). Evidence suggests that cerebellar LTD is mediated by a lasting desensitization of the AMPA receptor subtype. Although induction of cerebellar LTD requires coactivation of parallel and climbing fibers, the long-term depression of synaptic transmission is specific for the parallel fiber synapse with no effect on the climbing fiber synapse. Although empirical evidence for this form of synaptic plasticity in the cerebellum was not reported until 1982, theoretical models developed more than a decade prior to that predicted that transmission at the parallel fiber–Purkinje cell synapse could be modified by simultaneous activation of the climbing fiber input, and it was postulated that this lasting depression of synaptic transmission plays an important role in various forms of motor learning (Marr, 1969; Albus, 1971).

Although it is clear that activation of mGluRs alone is not sufficient for induction of LTD (Crepel et al., 1991; Glaum et al., 1992), several studies suggest that mGluRs participate in this response (Ito and Karachot, 1990; Linden et al., 1991; Daniel et al., 1992; *see* Chapter 7). As discussed above, stimulation of climbing fibers results in a large complex spike in Purkinje cells that is accompanied by a large increase in intracellular calcium that enters through voltage-dependent calcium channels. Stimulation of parallel fibers induces much smaller EPSPs, but the studies discussed above suggest that mGluRs are activated by activity at the parallel fiber synapse. Interestingly, stimulation of climbing fibers can induce LTD in the absence of parallel fiber stimulation if climbing fibers are stimulated in the presence of 1S,3R-ACPD. Furthermore, 1S,3R-ACPD can induce LTD in the absence of climbing fiber stimulation if added during simultaneous depolarization of Purkinje cells and induction of calcium spikes (Daniel et al., 1992). In the absence of pharmacological manipulations, activation of parallel fibers and climbing fibers may be required for activation of mGluRs and calcium-channel activity, respectively. This would explain the need for simultaneous activation of these two inputs for induction of LTD under normal conditions.

4. Roles of mGluRs in Regulation of Cardiovascular Function

4.1. Circuitry of the Baroreceptor Reflex

The most well-characterized mechanism for central control of arterial pressure is the baroreceptor reflex. This reflex is a negative feedback system that responds to an increase in arterial pressure with a decrease in sympathetic outflow and consequent bradycardia and decreased arterial pressure. Figure 3 gives a simplified diagram of the basic circuitry involved in the central baroreceptor reflex controlling sympathetic outflow (*see* Gordon and Talman, 1992 for review). A rise in arterial pressure activates stretch receptors (called baroreceptors or pressoreceptors) that are located in the walls of a number of large systemic arteries, but are most abundant in the carotid sinus and the wall of the aortic arch. Baroreceptor afferents project to the nucleus of the tractus solitarius (NTS), which lies in the medullary area of the brainstem. Excitatory afferents from the NTS project to the caudal ventrolateral medulla (CVM), which in turn sends an inhibitory projection to a functional area known as the "vasomotor center"

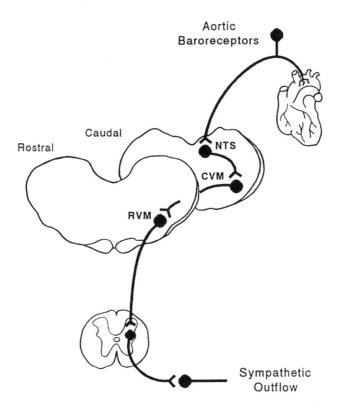

Fig. 3. Schematic illustration of central baroreflex pathways controlling sympathetic outflow. NTS = nucleus tractus solitarius, CVM = caudal ventrolateral medulla, RVM = rostral ventrolateral medulla.

in the rostral ventrolateral medulla (RVM). Afferents from the RVM project to the spinal cord, where they excite vasoconstrictor sympathetic preganglionic neurons. Thus, the net effect of activation of the baroreflex is reduction of sympathetic outflow and a consequent reduction of heart rate and arterial pressure.

Although this simplified outline of the major circuitry involved in the baroreceptor reflex is useful for discussing the roles of mGluRs in regulating cardiovascular function, it should be kept in mind that the detailed circuitry of each structure within this overall circuit is quite complex. For instance, the NTS has a complex intrinsic circuitry with a variety of neuronal types, receives input from other visceral afferents, as well as a variety of other brain regions, and projects to regions other than the RVM.

4.2. mGluRs May Participate in Depressor Responses by Actions in the NTS

Accumulating evidence suggests that glutamate is the neurotransmitter responsible for transmission of baroreceptor information from the baroreceptors to neurons in the NTS. Injection of glutamate, kainate, NMDA, quisqualate, or AMPA into the NTS induces cardiovascular responses that mimic the baroreflex response (Leone and Gordon, 1989; Talman, 1989) and selective ionotropic glutamate receptor antagonists inhibit the baroreceptor reflex (Leone and Gordon, 1989; Talman, 1989), suggesting that ionotropic glutamate receptors play a predominant role in transmission of the baroreflex in the NTS. However, although ionotropic glutamate receptor antagonists inhibit the baroreceptor reflex and inhibit cardiovascular responses to microinjection of selective ionotropic glutamate receptor agonists into the NTS, they do not inhibit the response to microinjection of the putative neurotransmitter glutamate (Leone and Gordon, 1989). Likewise, ionotropic glutamate receptor antagonists fail to block the cardiovascular response to microinjections of quisqualate (Talman, 1989). This prompted Pawloski-Dahm and Gordon (1992) to test the hypothesis that activation of mGluRs in the NTS can induce cardiovascular responses similar to those elicited by activation of the baroreceptor reflex. These workers found that microinjection of *trans*-ACPD into the NTS induces cardiovascular responses similar to those induced by other EAAs. As with the effects of glutamate and quisqualate, the cardiovascular effects of *trans*-ACPD are not blocked by the ionotropic glutamate receptor antagonist kynurenate. Thus, it is likely that mGluRs exist on neurons in the NTS, where they may play a role in regulating cardiovascular function.

Studies in NTS slices reveal that 1S,3R-ACPD directly depolarizes NTS neurons (Glaum and Miller, 1992). In addition, mGluR activation reduces both excitatory and inhibitory transmission in the

NTS (Glaum and Miller, 1992, 1993). 1S,3R-ACPD virtually abolishes evoked monosynaptic IPSPs and IPSCs in NTS neurons, an effect that is at least partially mediated by a postsynaptic reduction in $GABA_A$ receptor currents (Glaum and Miller, 1993). Both the direct excitatory effects on NTS neurons and the disinhibition induced by mGluR activation would be expected to increase the net excitatory output from the NTS and likely contribute to the cardiovascular responses to microinjection of mGluR agonists into this structure. Furthermore, these responses to 1S,3R-ACPD can be mimicked by high-frequency stimulation of the tractus solitarius (TS) in the presence of ionotropic glutamate receptor antagonists (Glaum and Miller, 1993). Although it is possible that the effects of TS stimulation are mediated by an unknown neurotransmitter, rather than activation of mGluRs, these data are consistent with the hypothesis that mGluRs on NTS neurons can be synaptically activated. However, as discussed above, ionotropic glutamate receptor antagonists completely block the baroreceptor reflex, suggesting that mGluRs are not activated at synapses involved in this reflex under normal physiological conditions. It is possible that the mGluRs are present at different synapses than those involved in the baroreceptor reflex and play a role in some other aspect of cardiovascular regulation. Alternatively, since high-frequency stimulation of the TS is required for mGluR activation, it may be that mGluRs are only activated in response to unusually intense activation of the baroreceptor afferents that were not achieved in the studies showing complete blockade of the baroreceptor reflex by ionotropic glutamate receptor antagonists. Further studies will be needed to determine the exact role that mGluRs in the NTS play in regulating cardiovascular function.

4.3. Activation of mGluRs in the RVM and CVM Has Opposite Effects on Cardiovascular Function

As can be appreciated from Fig. 3, because projections from the CVM to the RVM are inhibitory, increased output from the CVM would be expected to reduce sympathetic output, whereas increased output from the RVM would be expected to increase sympathetic output. Tsuchihashi and Averill (1993) reported evidence that activation of mGluRs has net excitatory effects on neurons in both the CVM and the RVM. Thus, microinjection of 1S,3R-ACPD into the CVM reduces sympathetic nerve activity and mean arterial pressure, whereas 1S,3R-ACPD microinjection into the RVM increases arterial pressure and sympathetic nerve activity. As with the effects of 1S,3R-ACPD in the NTS, these effects are not blocked by selective ionotropic glutamate receptor antagonists and are mimicked by microinjection of glutamate in the presence of ionotropic glutamate receptor antagonists.

5. Roles of mGluRs in Regulating Striatal Function

The basal ganglia are a highly interconnected set of forebrain and midbrain nuclei that play a critical role in generation of voluntary movement (*see* Wilson, 1990 for review). The caudate nucleus, putamen, and nucleus accumbens are three major structures of the basal ganglia that are very similar in their internal structure and together are referred to as the neostriatum. Other major structures of the basal ganglia include the substantia nigra (which consists of the pars reticulata and the pars compacta) and the globus pallidus.

5.1. Circuitry of the Basal Ganglia

The intrinsic circuitry of the basal ganglia and the connections of the basal ganglia with outside structures are highly complex and are beyond the scope of this chapter. However, Fig. 4 shows some of the major connections of the basal ganglia with a focus on the striatum. The major inputs into the basal ganglia enter at the level of the striatum. These inputs come from four major sources. The cerebral cortex and intralaminar nuclei of the thalamus send excitatory afferents to the striatum that are likely to involve glutamate or another excitatory amino acid as a transmitter. The dorsal raphe nucleus sends excitatory serotonergic projections to the striatum. The striatum also receives a major input from dopaminergic neurons in the pars compacta of the substantia nigra. The principal neurons in the striatum are the spiny neurons, which are GABAergic neurons that send major inhibitory projections to the globus pallidus and the

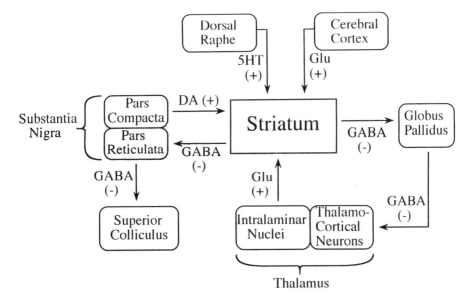

Fig. 4. Diagram showing some of the major connections of the striatum.
5HT = 5-hydroxytryptamine (serotonin), Glu = glutamate, DA = dopamine.
Excitatory and inhibitory projections are denoted by (+) and (–), respectively.

pars reticulata of the substantia nigra. Cells within the globus pallidus
and pars reticulata are also inhibitory and project to thalamocortical
neurons and superiour colliculus, respectively, where they exert a
tonic inhibitory influence.

An understanding of the function of the basal ganglia requires
a knowledge of the normal firing patterns of the principal neurons
within the stiatum and its target structures. The spiny neurons of the
striatum are generally electrically silent and do not fire spontaneous
action potentials. However, occasionally the cells fire a brief (0.1–2
s) train of action potentials. In contrast, the target cells of the striatal
spiny neurons in the globus pallidus and substantia nigra pars reticulata are tonically active with very high rates of firing. As mentioned
above, the cells within the globus pallidus and pars reticulata are
inhibitory, and this high firing rate exerts a tonic inhibition of their
target cells in the thalamus and superior colliculus. The brief bursts
of firing of striatal spiny neurons causes a pause in firing of globus

pallidus and pars reticulata neurons, and thereby disinhibits cells in the thalamus and superior colliculus. This increase in thalamic and superior colliculus cell firing ultimately leads to an increase in locomotor activity. Thus, in behaving animals, episodes of increased cell firing in the striatum and the corresponding decrease in cell firing in the globus pallidus have been shown to correspond to the onset of voluntary movements. Both the dopaminergic inputs from the substantia nigra pars compacta and the EAA inputs from cerebral cortex and thalamus can increase the net output of the striatum and thereby increase locomotor activity.

5.2. Localization of mGluRs Within the Striatum

Although little is known about the subcellular localization of mGluR subtypes within the striatum, Northern blot analysis and *in situ* hybridization studies reveal that mGluR1 (Shigemoto et al., 1992), mGluR2 (Ohishi et al., 1993b), mGluR4 (Kristensen et al. 1993), and mGluR5 (Abe et al., 1992) are expressed in the striatum, with mGluR5 showing particularly high expression (*see* Chapter 4). Furthermore, biochemical studies suggest that mGluRs in the striatum are coupled to a variety of second-messenger systems, including activation of phosphoinositide hydrolysis (Sladeczek et al., 1985; Desai and Conn, 1990), potentiation of cAMP responses (Winder et al., 1993), activation of phospholipase A2 (Dumuis et al., 1990, 1993), and inhibition of adenylyl cyclase (Prezeau et al., 1992).

5.3. Intrastriatal Injection of 1S,3R-ACPD Induces a Dopamine-Dependent Increase in Motor Output

As discussed above, evidence suggests that activity of dopaminergic inputs to the striatum induces a net increase in striatal output and a corresponding increase in locomotor activity. Selective unilateral lesions of the dopaminergic nigrostriatal pathway induces well-characterized behavioral effects that include postural changes with movements in the direction of the lesioned side of the brain. If selective dopamine receptor agonists are administered to rats with unilateral lesions of the nigrostriatal pathway, the animals respond with spontaneous repetitive rotations in a direction contralateral to the side of the lesion (Ungerstedt, 1971). This effect of dopamine agonists is thought to be the result of a supersensitivity of striatal dopamine receptors on the lesioned side of the brain that results in an imbalance in the effects of dopamine receptor agonists on striatal neurons in the two hemispheres.

Several studies suggest that glutamate released from the cortical inputs to the striatum can facilitate release of dopamine from presynaptic nerve endings in the striatum. This prompted Sacaan et al. (1991) to test the hypothesis that mGluR activation might enhance dopaminergic transmission in the striatum and induce contralateral turning in rats. Interestingly, these workers found that intrastriatal injection of 1S,3R-ACPD induced robust turning behavior in a direction contralateral to the injection site. This effect was not mimicked by NMDA or 1R,3S-ACPD (which is inactive at most mGluRs), or blocked by a selective NMDA receptor antagonist. L-AP3, which acts as a partial agonist at some mGluRs, induced a slight increase in contralateral turning and partially blocked the turning response to 1S,3R-ACPD. These data are consistent with the hypothesis that the effects of 1S,3R-ACPD are mediated by activation of mGluRs.

Although glutamate can facilitate dopaminergic transmission in the striatum, the mGluR-mediated increases in turning behavior could be mediated by direct activation of striatal spiny neurons, rather than potentiation of effects of the nigrostriatal input. However, Sacaan et al. (1992) showed that dopamine depletion with α-methyl-ρ-tyrosine resulted in an approx 85% inhibition of 1S,3R-ACPD-induced contralateral turning, and the dopamine receptor antagonist haloperidol inhibited the response by approx 50%. Furthermore, 1S,3R-ACPD increased the levels of dopamine metabolites in the striatum, and the time-course of this effect was closely associated with the time course of 1S,3R-ACPD-induced increases in contralateral turning. Taken together, these data suggest that this effect of 1S,3R-ACPD is at least partially mediated by facilitation of transmission at the dopaminergic synapses. However, direct effects of mGluR activation on striatal spiny neurons could also contribute to this effect. For instance, *trans*-ACPD induces inward currents and depolarization of striatal neurons, and may indirectly increase excitability of these cells by reducing synaptic inhibition (Calabresi et al., 1992). Another prominent action of *trans*-ACPD in the striatum is a marked reduction of transmission at glutamatergic synapses, an effect that is likely to be mediated by activation of presynaptic autoreceptors (Lovinger, 1991; Lovinger et al., 1993). Thus, mGluR activation may serve to enhance transmission at dopaminergic inputs to the striatum while simultaneously reducing transmission at cortico-striatal and/or thalamo-striatal synapses.

6. An AP4-Sensitive mGluR Is Important for Retinal Signal Transduction

Another brain region where a role for mGluRs has been established is in the retina. The retina has a well-characterized and complex circuitry that is discussed in detail by Sterling (1990). Information flows from photoreceptor cells to retinal interneurons, the bipolar cells, and finally to the output cells of the retina, the retinal ganglion cells. Neither photoreceptor cells nor bipolar cells are capable of producing action potentials, so information is passed on by passive spread of potentials in these cells. The retinal ganglion cells are capable of firing action potentials, and send projections to the lateral geniculate nucleus and the superior colliculus. Photoreceptor cells (rods and cones) are maintained in a depolarized state in the absence of light, and thus release their neurotransmitter, glutamate, in a constitutive manner. This depolarization is the result of the constitutive activation of a cGMP-gated cation channel. Light striking the photoreceptor cell leads a decrease in intracellular cGMP and resultant closure of the cGMP-gated channels. Closing these channels results in hyperpolarization of the photoreceptor cell and a consequent decrease in glutamate release from the terminals. The photoreceptor cells synapse onto two main classes of bipolar cells: ON and OFF. The OFF bipolar cells respond to glutamate with the stereotypical depolarization, whereas the ON bipolar cells respond to glutamate by hyperpolarizing. Thus, the effect of light shining on a photoreceptor cell is hyperpolarization of the OFF bipolar cell and depolarization of the ON bipolar cells.

As mentioned above, glutamate, which is tonically released from photoreceptor cells in the absence of light, depolarizes the OFF-bipolar cells and hyperpolarizes the ON-bipolar cells. How can glutamate exert these diametrically opposed actions on two very similar cell types? The answer lies in the types of receptors localized on the postsynaptic cells. The primary receptor type localized on OFF-bipolar cells is of the AMPA/Kainate subtype (Bloomfield and Dowling, 1985). In contrast, evidence suggests that the receptor type localized on the ON-bipolar cells is an L-AP4-sensitive mGluR (Nawy and Jahr, 1990; Shiells and Falk, 1992; Yamashita and Wassle, 1991). Activation of this receptor by either glutamate or L-AP4 inhibits a cGMP-activated cation current in retinal bipolar cells that is similar to that found in photoreceptor cells described above (Shiells and Falk, 1992). This mGluR inhibits the cGMP-gated current by a mechanism that depends on the function of pertussis toxin-sensitive G-proteins (Nawy and Jahr, 1990; Shiells and Falk, 1992; Kikkawa et al., 1993). This effect is likely to be mediated by inhibition of guanylyl cyclase or by activation of a cGMP phosphodiesterase.

Interestingly, Nakanishi and coworkers recently characterized a retina-specific mGluR (mGluR6) with high agonist-selectivity for L-AP4 and L-serine-O-phosphate (Nakajima et al., 1993). This clone was isolated from a rat retinal cDNA library by crosshybridization with a previously isolated cDNA clone for an mGluR. mGluR6 is highly homologous to the previously cloned mGluRs, and appears to be highly specific to retina based on Northern analysis and *in situ* hybridization (Nakajima et al, 1993). Specifically, mRNA for mGluR6 appears to be selectively expressed in the inner nuclear layer of the retina, where the ON-bipolar cells reside. mGluR6 couples to a pertussis-toxin-sensitive G-protein when expressed in Chinese hamster ovary cells. This clone shares its highest degree of homology with mGluR4, which is also activated well by L-AP4 and L-serine-O-phosphate (Nakajima et al., 1993). Based on the pharmacological properties of mGluR6 coupled with its high localization in the inner nuclear layer of the retina, it is likely that this mGluR mediates glutamate-induced hyperpolarization of ON-bipolar cells. If this proves to be the case, this may be one of the first physiological functions ascribed to a specific subtype of the cloned mGluRs.

7. Conclusions

In summary, although rapid progress is being made in our understanding of the molecular structure of mGluRs and the physiological effects of mGluR activation at a cellular level, we have only begun to gain an appreciation of the roles of mGluRs on transmission through intact neuronal circuits and regulation of animal behavior. However, because of the widespread distribution of glutamatergic neurons within the central nervous system, it is likely that mGluRs will ultimately be found to participate at some level in regulation of transmission through virtually every neural circuit in the brain. Indeed, although we focused on a limited number of neuronal circuits where the effects of mGluR activation have been particularly well characterized, mGluR agonists have been shown to modulate synaptic transmission and/or neuronal excitability in a wide variety of other brain regions. These include neocortex (Greene et al., 1992; Sayer et al., 1992; Wang and McCormick, 1993; Kato, 1993), spinal cord (McLennan and Liu, 1982; Pook et al., 1992; Bleakman et al., 1992; Cerne and Randic, 1992), thalamus (Salt and Eaton, 1991; McCormick and von Krosigk, 1992), septal nucleus (Zheng and Gallagher, 1991, 1992a,b), olfactory cortex (Constanti and Libri, 1992; Collins, 1993), and amygdala (Rainnie and Shinnick-Gallagher, 1992). A complete understanding of the roles of synaptically activated mGluRs in transmission through intact neural circuits in each of these regions must await development of specific mGluR antagonists. However, careful study of the physiological effects of exogenously applied mGluR agonists, coupled with an understanding of the anatomical distribution of glutamatergic synapses is rapidly shedding light on the roles of mGluRs in a variety of important functions of the CNS.

Acknowledgments

Work in the authors' laboratory is supported by NIH grants NS-28405 and NS-31373, a grant from the Council for Tobacco Research (PJC), an NIH NRSA predoctoral fellowship (DGW), and a Howard Hughes predoctoral fellowship (RWG).

References

Abe, T., Sugihara, H., Nawa, H., Shigemoto, R., Mizuno, N., and Nakanishi, S. (1992) Molecular characterization of a novel metabotropic glutamate receptor mGluR5 coupled to inositol phosphate/Ca2+ signal transduction. *J. Biol. Chem.* **267,** 13,361–13,368.

Albus, J. S. (1971) A theory of cerebellar function. *Math. Biosci.* **10,** 25–61.

Aniksztejn, L., Otani, S., and Ben-Ari, Y. (1992) Quisqualate metabotropic receptors modulate NMDA currents and facilitate induction of long-term potentiation through protein kinase C. *Eur. J. Neurosci.* **4,** 500–505.

Bashir, Z. I., Bortolotto, Z. A., Davies, C. H., Berretta, N., Irving, A. J., Seal, A. J., Henley, J. M., Jane, D. E., Watkins, J. C., and Collingridge, G. L. (1993) Induction of LTP in the hippocampus needs synaptic activation of glutamate metabotropic receptors. *Nature* **363,** 347–350.

Baskys, A. and Malenka, R. C. (1991a) Agonists at metabotropic glutamate receptors presynaptically inhibit EPSCs in neuronal rat hippocampus. *J. Physiol.* **444,** 687–701.

Baskys, A. and Malenka, R. C. (1991b) *Trans*-ACPD depresses synaptic transmission in the hippocampus. *Eur. J. Pharmacol.* **193,** 131,132.

Baskys, A., Wang, S., and Wojtowicz, J. M. (1993) Metabotropic agonist-induced changes in elementary synaptic events in the dentate gyrus neurons of the hippocampus. *Functional Neurol.* **8,** 9.

Batchelor, A. M. and Garthwaite, J. (1993) Novel synaptic potentials in cerebellar purkinje cells: probable mediation by metabotropic glutamate receptors. *Neuropharmacol.* **32,** 11–20.

Blackstone, C. D., Supattapone, S., and Snyder, S. H. (1989) Inositol phospholipid-linked glutamate receptors mediate cerebellar parallel-fiber-purkinje-cell synaptic transmission. *Proc. Natl. Acad. Sci. USA* **86,** 4316–4320.

Bleakman, D., Rusin, K. I., Chard, P. S., Glaum, S. R., and Miller, R. J. (1992) Metabotropic glutamate receptors potentiate ionotropic glutamate responses in the rat dorsal horn. *Mol. Pharmacol.* **42,** 192–196.

Bloomfield S. A. and Dowling J. E. (1985) Roles of aspartate and glutamate in synaptic transmission in rabbit retina. I. Outer plexiform layer. *J. Neurophysiol.* **53,** 699–713.

Boss, V. and Conn, P. J. (1992) Metabotropic excitatory amino acid receptor activation stimulates phospholipase D in hippocampal slices. *J. Neurochem.* **59,** 2340–2343.

Boss, V., Nutt, K. M., and Conn, P. J. (1994) L-Cysteine sulfinic acid as an endogenous agonist of a novel metabotropic receptor coupled to stimulation of phospholipase D activity. *Mol. Pharmacol.* (in press).

Brown, T. H. and Zador, A. M. (1990) Hippocampus, in *The Synaptic Organization of the Brain,* (Shepherd, G. M., eds.) Oxford University Press, New York, pp. 346–388.

Calabresi, P., Mercuri, N. B., and Bernardi, G. (1992) Activation of quisqualate metabotropic receptors glutamate and GABA-mediated synaptic potentials in the rat striatum. *Neurosci. Lett.* **139,** 41–44.

Cerne, R. and Randic, M. (1992) Modulation of AMPA and NMDA responses in rat spinal dorsal horn neurons by *trans*-1-aminocyclopentane-1,3-dicarboxylic acid. *Neurosci. Lett.* **144,** 180–184.

Charpak, S. and Gahwiler, B. (1991) Glutamate mediates a slow synaptic response in hippocampal slice cultures. *Proc. R. Soc. Lond.* **243,** 221–226.

Charpak, S., Gahwiler, B. H., Do, K. Q., and Knopfel, T. (1990) Potassium conductances in hippocampal neurons blocked by excitatory amino-acid transmitters. *Nature* **347,** 765–767.

Collins, G. G. S. (1993) Actions of agonists of metabotropic glutamate receptors on synaptic transmission and transmitter release in the olfactory cortex. *Br. J. Pharmacol.* **108,** 422–430.

Constanti, A. and Libri, V. (1992) Trans-ACPD induces a slow post-stimulus inward tail current (I_{ADP}) in guinea-pig olfactory cortex neurones in vitro. *Eur. J. Pharmacol.* **214,** 105,106.

Crepel, F., Daniel, H., Hemart, N., and Jaillard, D. (1991) Effects of ACPD and AP3 on parallel-fibre-mediated EPSPs of purkinje cells in cerebellar slices in vitro. *Exp. Brain Res.* **86,** 402–406.

Daniel, H., Hemart, N., Jaillard, D., and Crepel, F. (1992) Coactivation of metabotropic glutamate receptors and of voltage-gated calcium channels induces long-term depression in cerebellar purkinje cells in vitro. *Exp. Brain Res.* **90,** 327–331.

Desai, M. A. and Conn, P. J. (1990) Selective activation of phosphoinositide hydrolysis by a rigid analogue of glutamate. *Neurosci. Lett.* **109,** 157–162.

Desai, M. A. and Conn, P. J. (1991) Excitatory effects of ACPD receptor activation in the hippocampus are mediated by direct effects on pyramidal cells and blockade of synaptic inhibition. *J. Neurophysiol.* **66,** 40–52.

Desai, M. A. and Conn, P. J. (1992) Activation of metabotropic glutamate receptors decreases synaptic inhibition in hippocampus by reducing excitation of inhibitory interneurons. *Soc. Neurosci. Abs.* **18,** 804.

Desai, M. A., Smith, T. S., and Conn, P. J. (1992) Multiple metabotropic glutamate receptors regulate hippocampal function. *Synapse* **12,** 206–213.

Doze, V. A., Cohen, G. A., and Madison, D. V. (1991) Synaptic localization of adrenergic disinhibition in the rat hippocampus. *Neuron* **6,** 889–900.

Dumuis, A., Pin, J. P., Oomagari, K., Sebben, M., and Bockaert, J. (1990) Arachidonic acid released from striatal neurons by joint stimulation of ionotropic and metabotropic quisqualate receptors. *Nature* **347,** 181–183.

Dumuis, A., Sebben, M., Fagni, L., Prezeau, L., Manzoni, O., Cragoe, E. J., Jr., and Bockaert, J. (1993) Stimulation by glutamate receptors of arachidonic acid release depends on the Na^+/Ca^{2+} exchanger in neuronal cells. *Mol. Pharmacol.* **43,** 976–981.

East, S. J. and Garthwaite, J. (1992) Actions of a metabotropic glutamate receptor agonist in immature and adult rat cerebellum. *Eur. J. Pharmacol.* **219,** 395–400.

Fagni, L., Bossu, J. L., and Bockaert, J. (1991) Activation of a large-conductance Ca^{2+}-dependent K^+ channel by stimulation of glutamate phosphoinositide-coupled receptors in cultured cerebellar granule cells. *Eur. J. Neurosci.* **3,** 778–789.

Fotuhi, M., Sharp, A. H., Glatt, C. E., Hwang, P. M., Krosigk, M. V., Snyder, S. H., and Dawson, T. M. (1993) Differential localization of phosphoinositide-linked metabotropic glutamate receptor (mGluR1) and the inositol 1,4,5,-trisphosphate receptor in rat brain. *J. Neurosci.* **13(5)**, 2001–2012.

Genazzani, A. A., Casabona, G., L'Episcopo, M. R., Condorelli, D. F., Dell'Albani, P., Shinozaki, H., and Nicoletti, F. (1993) Characterization of metabotropic glutamate receptors negatively linked to adenylyl cyclase in brain slices. *Brain Res.* **622**, 132–138.

Gereau, R. W. and Conn, P. J. (1993) A cyclic AMP-dependent form of associative synaptic plasticity induced by coactivation of β-adrenergic receptors and metabotropic glutamate receptors in rat hippocampus. *J. Neurosci.* **14**, 3310–3318.

Ghez, C. (1991) The Cerebellum, in *Principles of Neuroscience* (Kandel, E. R., Schwartz, J. H., and Jessell, T. M., eds.), Elsevier, New York, pp. 626–646.

Glaum, S. R. and Miller, R. J. (1992) Metabotropic glutamate receptors mediate excitatory transmission in the nucleus of the solitary tract. *J. Neurosci.* **12**, 2251–2258.

Glaum, S. R. and Miller, R. J. (1993) Activation of metabotropic glutamate receptor produces reciprocal regulation of ionotropic glutamate and GABA responses in the nucleus of the tractus solitarius of the rat. *J. Neurosci.* **13(4)**, 1636–1641.

Glaum, S. R., Slater, N. T., Rossi, D. J., and Miller, R. J. (1992) Role of metabotropic glutamate (ACPD) receptors at the parallel fiber-purkinje cell synapse. *Eur. J. Pharmacol.* **219**, 395–400.

Gordon, F. J. and Talman, W. T. (1992) Role of excitatory amino acids and their receptors in bulbospinal control of cardiovascular function, in *Central Neural Mechanisms in Cardiovascular Regulation,* vol. 2 (Kunos, G. and Ciriello, J., eds.) Birkhauser, Boston, pp. 209–225.

Greene, C., Schwindt, P., and Crill, W. (1992) Metabotropic receptor mediated after depolarization in neocortical neurons. *Eur. J. Pharmacol.—Mol. Pharmacol.* **226**, 279,280.

Houamed, K. M., Kuijper, J. L., Gilbert, T. L., Haldeman, B. A., O'Hara, P. J., Mulvihill, E. R., Almers, W., and Hagen, F. S. (1991) Cloning, expression, and gene structure of a G protein- coupled glutamate receptor from rat brain. *Science* **252**, 1318–1321.

Hu, G. and Storm, J. F. (1991) Excitatory amino acids acting on metabotropic glutamate receptors broaden the action potential in hippocampal neurons. *Brain Res.* **568**, 339–344.

Hu, G. and Storm, J. F. (1992) 2-Amino-3-phosphonopropionate fails to block postsynaptic effects of metabotropic glutamate receptors in rat hippocampal neurones. *Acta Physiol. Scand.* **145**, 187–191.

Ito, M. (1989) Long-term depression. *Annual Rev. Neurosci.* **12**, 85–102.

Ito, M. and Karachot, L. (1990) Receptor subtypes involved in, and time course of, the long-term desensitization of glutamate receptors in cerebellar purkinje cells. *Neurosci. Res.* **8**, 303–307.

Kato, N. (1993) Dependence of long-term depression on postsynaptic metabotropic glutamate receptors in visual cortex. *Proc. Natl. Acad. Sci. USA* **90**, 3650–3654.

Kikkawa, S., Nakagawa, M., Iwasa, T., Kaneko, A., and Tsuda, M. (1993) GTP-binding protein couples with metabotropic glutamate receptor in bovine retinal on-bipolar cell. *Biochem. Biophys. Res. Commun.* **195**, 374–379.

Kristensen, P., Suzdak, P. D., and Thomsen, C. (1993) Expression pattern and pharmacology of the rat type IV metabotropic glutamate receptor. *Neurosci. Lett.* **155,** 159–162.

Leone, C. and Gordon, F. J. (1989) Is L-glutamate a neurotransmitter of baroreceptor information in the nucleus of the tractus solitarius?. *J. Pharmacol. Exp. Ther.* **250,** 953–962.

Linden, D. J., Dickinson, M. H., Smeyne, M., and Connor, J. A. (1991) A long-term depression of AMPA currents in cultured cerebellar purkinje neurons. *Neuron* **7,** 81–89.

Littman, L., Glatt, B. S., and Robinson, M. B. (1993) Multiple subtypes of excitatory amino acid receptors coupled to the hydrolysis of phosphoinositides in rat brain. *J. Neurochem.* **61,** 586–593.

Liu, Y., Disterhoft, J. F., and Slater, N. T. (1993) Activation of metabotropic glutamate receptors induces long-term depression of GABAergic inhibition in hippocampus. *J. Neurophysiol.* **69,** 1000–1004.

Llinas, R. R. and Walton, K. D. (1990) Cerebellum, in *The Synaptic Organization of the Brain,* (Anonymous, ed.) Oxford University Press, New York, pp. 214–246.

Lothman, E. W., Bertram, E. H. I., and Stringer, J. L. (1991) Functional anatomy of hippocampal seizures. *Prog. Neurobiol.* 37, 1–82.

Lovinger, D. M. (1991) *Trans*-1-aminocyclopentante-1,3-dicarboxylic acid (*t*-ACPD) decreases synaptic excitation in rat striatal slices through a presynaptic action. *Neurosci. Lett.* **129,** 17–21.

Lovinger, D. M., Tyler, E., Fidler, S., and Merritt, A. (1993) Properties of a presynaptic metabotropic glutamate receptor in rat neostriatal slices. *J. Neurophysiol.* **69,** 1236–1244.

Madison, D. V. and Nicoll, R. A. (1986) Actions of noradrenaline recorded intracellularly in rat hippocampal CA1 pyramidal neurones, in vitro. *J. Physiol.* **372,** 221–224.

Madison, D. V. and Nicoll, R. A. (1988) Enkephalin hyperpolarizes interneurones in the rat hippocampus. *J. Physiol.* **398,** 123–130.

Marr, D. (1969) A theory of cerebellar cortex. *J. Physiol.* **202,** 437–470.

Martin, L. J., Blackstone, C. D., Huganir, R. L., and Price, D. L. (1992) Cellular localization of a metabotropic glutamate receptor in rat brain. *Neuron* **9,** 259–270.

Masu, M., Tanabe, Y., Tsuchida, K., Shigemoto, R., and Nakanishi, S. (1991) Sequence and expression of a metabotropic glutamate receptor. *Nature* **349,** 760–765.

McCormick, D. A. and von Krosigk, M. (1992) Corticothalamic activation modulates thalamic firing through glutamate "metabotropic" receptors. *Proc. Natl. Acad. Sci. USA* **89,** 2774–2778.

McLennan, H. and Liu, J. (1982) The action of six antagonists of the excitatory amino acids on neurons of the rat spinal cord. *Exp. Brain Res.* **45,** 151–156.

Miles, R. and Poncer, J. (1993) Metabotropic glutamate receptors mediate a post-tetanic excitation of guinea-pig hippocampal inhibitory neurones. *J. Physiol.* **463,** 461–473.

Nakajima, Y., Iwakabe, H., Akazawa, C., Nawa, H., Shigemoto, R., Mizuno, N., and Nakanishi, S. (1993) Molecular characterization of a novel retinal metabotropic glutamate receptor mGluR6 with a high agonist selectivity for L-2-amino-4-phosphonobutyrate. *J. Biol. Chem.* **268**, 11,868–11,873.

Nawy, S. and Jahr, C. E. (1990) Suppression by glutamate of cGMP-activated conductances in retinal bipolar cells. *Nature* **346**, 269–271.

Nicoletti, F., Iadarola, M. F., Wroblewski, J. T., and Costa, E. (1986) Excitatory amino acid recognition sites coupled with inositol phospholipid metabolism: Developmental changes and interaction with α1-adrenoreceptors. *Proc. Natl. Acad. Sci. USA* **83**, 1931–1935.

Ohishi, H., Shigemoto, R., Nakanishi, S., and Mizuno, N. (1993a) Distribution of the mRNA for a metabotropic glutamate receptor (mGluR3) in the rat brain: An *in situ* hybridization study. *J. Comp. Neurol.* **335**, 252–266.

Ohishi, H., Shigemoto, R., Nakanishi, S., and Mizuno, N. (1993b) Distribution of the messenger RNA for a metabotropic glutamate receptor, mGluR2, in the central nervous system of the rat. *Neuroscience* **53**, 1009–1018.

Okada, D. (1992) Two pathways of cyclic GMP production through glutamate receptor-mediated nitric oxide synthesis. *J. Neurochem.* **59**, 1203–1210.

Okamoto, N., Hori, S., Akazawa, C., Hayashi, Y., Shigemoto, R., Mizuno, N., and Nakanishi, S. (1994) Molecular characterization of a new metabotropic glutamate receptor mGluR7 coupled to inhibitory cyclic AMP signal transduction. *J. Biol. Chem.* (in press).

Oleskevich, S. and Lacaille, J. (1992) Reduction of GABA$_b$ inhibitory postsynaptic potentials by serotonin via pre- and postsynaptic mechanisms in CA3 pyramidal cells of rat hippocampus in vitro. *Synapse* **12**, 173–188.

Pacelli, G. J. and Kelso, S. R. (1991) *Trans*-ACPD reduces multiple components of synaptic transmission in the rat hippocampus. *Neurosci. Lett.* **132**, 267–269.

Pawloski-Dahm, C. and Gordon, F. J. (1992) Evidence for a kynurenate-insensitive glutamate receptor in nucleus tractus solitarii. *Am. J. Physiol.* **262**, H1611–H1615.

Pin, J., Waeber, C., Prezeau, L., Bockaert, J., and Heinemann, S. F. (1992) Alternative splicing generates metabotropic glutamate receptors inducing different patterns of calcium release in *Xenopus* oocytes. *Proc. Natl. Acad. Sci. USA* **89**, 10,331–10,335.

Pook, P. C.-K., Sunter, D. C., Udvarhelyi, P. M., and Watkins, J. C. (1992) Evidence for presynaptic depression of monosynaptic excitation in neonatal rat motoneurones by (1S,3S)- and (1S,3R)-ACPD. *Exp. Physiol.* **77**, 529–532.

Prezeau, L., Manzoni, O., Homburger, V., Sladeczek, F., Curry, K., and Bockaert, J. (1992) Characterization of a metabotropic glutamate receptor: direct negative coupling to adenylyl cyclase and involvement of a pertussis toxin-sensitive G protein. *Proc. Natl. Acad. Sci. USA* **89**, 8040–8044.

Rainnie, D. G. and Shinnick-Gallagher, P. (1992) *Trans*-ACPD and L-APB presynaptically inhibit excitatory glutamatergic transmission in the basolateral amygdala (BLA). *Neurosci. Lett.* **139**, 87–91.

Sacaan, A. I. and Schoepp, D. D. (1992) Activation of hippocampal metabotropic excitatory amino acid receptors leads to seizures and neuronal damage. *Neurosci. Lett.* **139**, 77–81.

Sacaan, A. I., Bymaster, F. P., and Schoepp, D. D. (1992) Metabotropic glutamate receptor activation produces extrapyramidal motor system activation that is mediated by striatal dopamine. *J. Neurochem.* **59**, 245–251.

Sacaan, A. I., Monn, J. A., and Schoepp, D. D. (1991) Intrastriatal injection of a selective metabotropic excitatory amino acid receptor agonist induces contralateral turning in the rat. *J. Pharmacol. Exp. Ther.* **259,** 1366–1370.

Sah, P., Hestrin, S., and Nicoll, R. A. (1989) Tonic activation of NMDA receptors by ambient glutamate enhances excitability of neurons. *Science* **246,** 815–818.

Salt, T. E. and Eaton, S. A. (1991) Excitatory actions of the metabotropic excitatory amino acid receptor agonist, *trans*-(±)-1-amino-cyclopentane-1,3-dicarboxylate (*t*-ACPD), on rat thalamic neurons in vivo. *Eur. J. Neurosci.* **3,** 1104–1111.

Sayer, R. J., Schwindt, P. C., and Crill, W. E. (1992) Metabotropic glutamate receptor-mediated suppression of L-type calcium current in acutely isolated neorcortical neurons. *J. Neurophysiol.* **68,** 833–842.

Schoepp, D. D. and Conn, P. J. (1993) Metabotropic glutamate receptors in brain function and pathology. *Trends Pharmacol. Sci.* **14,** 13–20.

Shiells, R. A. and Falk, G. (1992) Properties of the cGMP-activated channel of retinal on-bipolar cells. *Proc. R. Soc. Lond.* **247,** 21–25.

Shigemoto, R., Nakanishi, S., and Mizuno, N. (1992) Distribution of the mRNA for a metabotropic glutamate receptor (mGluR1) in the central nervous system: an in situ hybridization study in adult and developing rat. *J. Comp. Neurol.* **322,** 121–135.

Sladeczek, F., Pin, J. P., Recasens, M., Bockaert, J., and Weiss, S. (1985) Glutamate stimulates inositol phosphate formation in striatal neurones. *Nature* **317,** 717,718.

Staub, C., Vranesic, I., and Knopfel, T. (1992) Responses to metabotropic glutamate receptor activation in cerebellar purkinje cells: Induction of an inward current. *Eur. J. Neurosci.* **4,** 832–839.

Sterling. P. (1990) Retina, in *The Synaptic Organization of the Brain,* (Shepherd, G. M., ed.) Oxford University Press, New York, pp. 170–213.

Stratton, K. R., Worley, P. F., and Baraban, J. M. (1989) Excitation of hippocampal neurons by stimulation of glutamate Q_p receptors. *Eur. J. Pharmacol.* **173,** 235–237.

Takagi, H., Takimizu, H., Barry, J., Kudo, Y., and Yoshioka, T. (1992) The expression of presynaptic t-ACPD receptor in rat cerebellum. *Biochem. Biophys. Res. Commun.* **189,** 1287–1295.

Talman, W. T. (1989) Kynurenic acid microinjected into the nucleus tractus solitarius of rat blocks the arterial baroreflex but not responses to glutamate. *Neurosci. Lett.* **102,** 247–252.

Tanabe, Y., Masu, M., Ishii, T., Shigemoto, R., and Nakanishi, S. (1992) A family of metabotropic glutamate receptors. *Neuron* **8,** 169–179.

Tanabe, Y., Nomura, A., Masu, M., Shigemotor, R., Mizuno, N., and Nakanishi, S. (1993) Signal transduction, pharmacological properties, and expression patterns of two rat metabotropic glutamate receptors, mGluR3 and mGluR4. *J. Neurosci.* **13,** 1372–1378.

Tsuchihashi, T. and Averill, D. B. (1993) Metabotropic glutamate receptors in the ventrolateral medulla of rats. *Hypertension* **21,** 739–744.

Ungerstedt, U. (1971) Postsynaptic supersensitivity after 6-hydroxydopamine induced degeneration of nigro-striatal dopamine system. *Acta Physiol. Scand. Suppl.* **367,** 69–93.

Vranesic, I., Staub, C., and Knopfel, T. (1993) Activation of metabotropic glutamate receptors induces an outward current which is potentiated by methylxanthines in rat cerebellar purkinje cells. *Neurosci. Res.* **16,** 209–215.

Wang, Z. and McCormick, D. A. (1993) Control of firing mode of corticotectal and corticopontine layer V burst-generating neurons by norepinephrine, acetylcholine, and 1S,3R-ACPD. *J. Neurosci.* **13(5),** 2199–2216.

Wilson, C. J. (1990) Basal ganglia, in The Synaptic Organization of the Brain, (Shepherd, G. M., ed.) Oxford University Press, New York, 279–316.

Winder, D. G. and Conn, P. J. (1993) Activation of metabotropic glutamate receptors increases cAMP accumulation in hippocampus by potentiating responses to endogenous adenosine. *J. Neurosci.* **13,** 38–44.

Winder, D. G., Smith, T. S., and Conn, P. J. (1993) Pharmacological differentiation of metabotropic glutamate receptors coupled to potentiation of cAMP responses and phosphoinositide hydrolysis. *J. Pharmacol. Exp. Ther.* **266,** 518–525.

Wroblewska, B., Wroblewski, J. T., Saab, O. H., and Neale, J. H. (1993) *N*-acetylaspartylglutamate inhibits forskolin-stimulated cyclic AMP levels via a metabotropic glutamate receptor in cultured cerebellar granule cells. *J. Neurochem.* **61,** 943–948.

Yamashita, M. and Wassle, H. (1991) Responses of rod bipolar cells from the rat retina to the glutamate agonist 2-amino-4-phosphonobutyric acid (APB). *J. Neurosci.* **11,** 2372–2382.

Zheng, F. and Gallagher, J. P. (1991) *Trans*-ACPD (*trans*-D,L-1-amino-1,3-cyclopentanedicarbioxylic acid elicited oscillation of membrane potentials in rat dorsolateral septal nucleus neurons recorded intracellularly in vitro. *Neurosci. Lett.* **125,** 147–150.

Zheng, F. and Gallagher, J. P. (1992a) Metabotropic glutamate receptors are required for the induction of long-term potentiation. *Neuron* **9,** 163–172.

Zheng, F. and Gallagher, J. P. (1992b) Burst firing of rat septal neurons induced by 1S,3R-ACPD requires influx of extracellular calcium. *Eur. J. Pharmacol.* **211,** 281,282.

CHAPTER 9

Role of Metabotropic Glutamate Receptors in Neuronal Degeneration

Jitendra Patel and William C. Zinkand

1. Introduction

Glutamate is believed to be the key mediator in neurodegenerative processes associated with stroke, epilepsy , and a broad spectrum of neurodegenerative disorders, including, Huntington's disease and Alzheimer disease (Choi, 1988; Beal, 1993). The elucidation of the underlying mechanism involved in the pathophysiology of glutamate has been the topic of intensive research. Although the role of glutamate-induced neurotoxicity in chronic neurodegenerative disease remains unclear, compelling evidence suggests that ischemia-triggered neuronal damage is largely attributed to excessive and persistent activation of glutamate receptors (for recent reviews, *see* Whetsell and Shapira, 1993; Choi, 1988; Beal, 1933). Because glutamate can produce both excitation and toxicity in neurons, the term "excitotoxicity" was proposed by Olney and coworkers (Olney et al., 1978). It is now widely believed that excessive elevation of

The Metabotropic Glutamate Receptors Eds.: P. J. Conn and J. Patel
© 1994 Humana Press Inc., Totowa, NJ

intracellular calcium is an early and critical step in excitotoxicity (*see* Fig 1; and Rothman, 1984; Choi, 1988; Garthwaite and Garthwaite, 1986; Frandsen and Schousboe, 1993). In vivo evidence supporting a central role of calcium in excitotoxicity is derived from the observation that calcium accumulates in nervous tissue in cerebral ischemia (Siesjo and Bengtsson, 1989; Simon et al., 1984) and in epilepsy (Meyer, 1989; Uematsu et al., 1990), and that ischemic cell damage can be attenuated by suppression of calcium influx (Pizzi et al., 1991; Valentino et al., 1993) and chelation of intracellular calcium (Tymianski et al., 1993).

Four distinct classes of glutamate receptor have been described. These are α-amino-3-hydroxy-5-methyl-4-isoxazolepropionic acid (AMPA), kainate, N-methyl-D-aspatate (NMDA), and metabotropic receptor subtypes. AMPA and kainate receptors mediate fast synaptic transmission through ionotropic channels, and can increase intracellular calcium either directly (Hoffmann et al., 1991) or via the activation of voltage-operated calcium channels (Mayer and Miller, 1991). The NMDA receptor channels are highly permeable to calcium and on activation lead to significant increase in intracellular Ca^{2+}. Certain metabotropic glutamate receptors (mGluRs), specifically mGluR1 and mGluR5, are coupled to phospholipase C and can therefore trigger mobilization of inositol trisphosphate (IP-3)-sensitive calcium stores (Murphy and Miller, 1988; Irving et al., 1992; Connor and Tseng, 1988; Frandsen and Schousboe, 1991). Since members of each glutamate receptor class can independently elicit an increase in intracellular Ca^{2+} (Mayer and Miller, 1991), it can be argued that each class could mediate glutamate excitotoxicity. It is generally regarded that during acute insults, such as stroke, hypoglycemia, trauma, and epilepsy, overstimulation of the NMDA receptor is a predominant mechanism for excessive entry of Ca^{2+} into neurons with subsequent damage (Choi, 1988; Meldrum and Gartwaite, 1990). However, recent demonstration that a potent and selective AMPA receptor antagonist, 2,3-dihydroxy-6-nitro-7-sulfamoyl-benzo(f)quinoxaline (NBQX), can greatly attenuate ischemia-induced neuronal loss has led some to suggest that the AMPA receptor may also play a role in the excitotoxicity process (Judge et al., 1991).

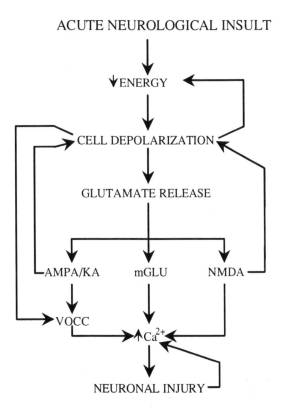

Fig. 1. Potential cascade of events contributing to excitoxicity. Key events appear to be excessive release of neurotransmitters, including glutamic acid in response to depolarization triggered by loss of energy, activation of glutamate receptors, and VOCC with attendant increase in calcium. In addition to contributing to the calcium overload, mGluR activation may exacerbate excitotoxicity by facilitation of glutamic acid release and potentiation of the ion-channel activities of the NMDA receptor and the VOCC. Alternatively, mGluR activation may suppress excitotoxicity by inhibition of glutamic acid release.

Unfortunately, because of the lack of selective pharmacological tools for the mGluRs, the role of this receptor subtype in excitotoxicity has been difficult to decipher. This notwithstanding, a number of recent studies, some using recently developed selective mGluR compounds, are beginning to shed new light on this subject.

2. Quisqualic Acid-Induced Neurotoxicity

Quisqualic acid (QA), which can activate mGluRs and AMPA receptors, is a potent neurotoxin in cortical cultures (Zinkand et al., 1992; Koh et al., 1990). The potency of QA neurotoxicity depended on the length of exposure; the EC50, for brief (5 min) and prolonged (20–24 h) exposure was, respectively, about 250 (Zinkand et al., 1992) and 1 μ*M* (Koh et al., 1990), respectively. Central injection of QA (Silverstein et al., 1986) or exposure of slices of young (8-d-old) rat hippocampus to QA (Garthwaite and Garthwaite, 1989b) also produced extensive cell death. In hippocampal slices and elsewhere (Zinkand et al., 1992; Frandsen and Schousboe, 1991), acute QA neurotoxicity was independent of extracellular calcium and was found to be insensitive to NMDA and non-NMDA antagonists. Interestingly, in slices, the pathological pattern of QA neurotoxicity, although distinct from that observed with NMDA, was similar to that observed in vivo following intense seizure activity (Evans et al., 1983) or central injection of glutamic acid (Sloviter and Dempster, 1985). The mechanism of QA neurotoxicity appears to involve a novel process. It seems QA is first taken up by neurons during initial incubation and is subsequently released gradually, converting a brief addition into prolonged exposure during which toxicity is triggered via the AMPA receptors (Zinkand et al., 1992; Koh et al., 1990; Garthwaite and Garthwaite, 1989b). Since the uptake of QA is insensitive to AMPA antagonists, treatment with these agents, if limited to brief exposure, is ineffective against the ensuing toxicity. However, prolonged treatment with AMPA antagonists can effectively block QA neurotoxicity. The mechanism of QA neurotoxicity described above may serve as an interesting illustration of how a brief presentation of an environmental toxin can lead to prolonged internal exposure and toxicity.

3. Neuroprotection by mGluR Agonists

Interestingly, QA at subtoxic concentration attenuates EAA neurotoxicity (McCaslin and Smith, 1988). The selective mGluR agonist (±) *trans*-1-aminocyclopentane-1,3-dicarboxylic and (*trans*-

ACPD) is also neuroprotective against EAA neurotoxicity. For example, it attenuated EAA-mediated neurotoxicity in cortical cultures (Koh et al., 1991), cerebellar granule cells (Pizzi et al., 1993), and retina (Birrell et al., 1993); protection against neurotoxicity induced by NMDA (Siliprandi et al., 1992; Lipartiti et al., 1993) and midcerebral artery occlusion (Chiamulera et al., 1992) has also been noted in vivo. In addition, *trans*-ACPD can prevent degeneration induced by low-potassium content media (10 m*M*) in cerebellar granule cells (Bruno et al., 1993). The neuroprotection by *trans*-ACPD could be reversed by protein kinase inhibitors, H-7 and HA-1004 (Koh et al., 1991), suggesting the involvement of a protein kinase in the neuroprotective action of *trans*-ACPD. However, *trans*-azetidine-2,4-dicarboxylic acid (trans-ADA), a novel agonist at phospholipase C-coupled mGluRs (Favaron et al., 1993), appears to be inactive against NMDA toxicity (Bruno et al., 1993). Moreover, recent observation that selective agonists at mGluR negatively coupled to adenylyl cyclase, L-CCG-1, DCG-4, and L-SOP (Bruno et al., 1993) can also attenuate NMDA neurotoxicity may suggest that the *trans*-ACPD neuroprotective effect is instead mediated by inhibition of adenylyl cyclase.

An alternative possibility could be that the neuroprotection is the result of inhibition of voltage-operated calcium channels (VOCC) (Sahara and Westbrook, 1993; Randall et al., 1993). Such action would result in a decrease in EAA-elicited Ca^{2+} increase and associated Ca^{2+}-toxicity (Schwartz et al., 1993). Consistent with this notion are number of reports demonstrating that blockade of VOCC can ameliorate excitotoxicity (Pizzi et al., 1991; Valentino et al., 1993), and that activation of other receptors that are negatively coupled to Ca^{2+}-channel and adenylyl cyclase, such as serotonin-1a (Prehn et al., 1991) and adenosine-A1 (Rudolphi et al.; 1992), are neuroprotective. Yet another possible mechanism for neuroprotective action of *trans*-ACPD could be via activation of certain K^+ currents (Fagni et al., 1991); by producing hyperpolarization of neurons, this should tend to reduce the efficacy of ionotropic EAA receptor-induced excitation and thus attenuate excitotoxicity. Finally, regardless of the second messenger involved, it seems that the neuroprotective action of 1S,3R-ACPD is mediated through a G-protein, since the neuro-protective effect can be prevented by pretreatment with pertussis toxin (Schwartz et al., 1993).

4. mGluRs and Cell Survival

Enhanced mGluR coupling to phospholipase C during early stages of development (Nicoletti et al., 1986; Dubek et al., 1989; Palmer et al., 1990) has led to the postulation that mGluRs may play an important role during postnatal development (*see* Chapter 10). Activity of phospholipase C-coupled mGluR during neuronal development and differentiation may serve to provide a general trophic signal, and thereby promote survival and confer neuroprotection from apoptotic cell death. If so, disruption or suppression of this mGluR action may impair development and cause widespread neurodegeneration. A recent illustration of this phenomena was noted by Bruno et al. (1993) in cerebellar granule cells. These investigators found that *trans*-ACPD prevented apoptotic cell death induced in these cultures by switching from the optimal depolarization media (media with 25 mM K$^+$) to a nondepolarizing low-K$^+$ media. Also, suppression of mGluR activity by chronic treatment of infant rodents with the mGluR antagonists AP-3 or ABH over several days was found to cause degeneration of virtually all nerve cells in the retina and some neurons in the brain (Olney et al., 1993; Fix et al., 1993). In cerebellar Purkinje cells cultures, blockade of ionotropic receptors increased cell number, suggesting that endogenous ionotropic activity decreased cell survival. By contrast, 1S,3R-ACPD decreased, and L-AP3 increased cell survival, again suggesting that endogenous mGluR activity is necessary for survival (Mount et al., 1993).

5. Neurotoxicity of mGluR Agonists: In Vivo Studies

Although in vitro, *trans*-ACPD displays little or no intrinsic toxicity, intrastriatal or hippocampal injection of 1S,3R-ACPD ($\geq$500 nmol), but not 1R-3S-ACPD, in 7-d-old neonatal rats produced dose-dependent brain injury. 1S,3R-ACPD toxicity was dose-dependent, occurred in the absence of prominent behavioral convulsions, and was qualitatively distinct from the NMDA-induced excitotoxicity (McDonald et al., 1993). 1S,3R-ACPD injury at 4-h postinjection, as revealed by light microscopy, was restricted to select neurons scattered throughout caudate-putamen and hippocampus, and produced significant losses of brain weight on the injected side 5 d postinjection. 1S,3R-ACPD-injury was insensitive to both NMDA and AMPA antagonists, but could be blocked by dantrolene (McDonald et al., 1993). Dantrolene inhibits Ca^{2+}-release from sarcoplasmic reticulum, and on this basis, the authors suggest that 1S,3R-ACPD toxicity could be mediated by a phospholipase C-coupled mGluR. This view is supported by recent reports demonstrating that 1S,3R-ACPD neurotoxicity is attenuated by the noncompetitive antagonist of the mGluR-mediated phosphoinositide metabolism, AP-3 (Olney et al., 1993; Lipartiti et al., 1993).

1S,3R-ACPD-induced injury appears to be age-specific. Although 1S,3R-ACPD toxicity can be observed in adult animals (Sacaan and Schoepp, 1992), the injury in adults may be mediated by a distinct mechanism, since the pathological pattern is different and the degeneration elicited can be attenuated by NMDA antagonists. The susceptibility of neonatal tissue to *trans*-ACPD toxicity may be conferred by enhanced mGluR coupling to phospholipase C during early stages of development (Nicoletti et al., 1986; Dubek et al., 1989; Palmer et al., 1990). The subcellular mechanism of *trans*-ACPD neurotoxicity is not clear. It is, however, possible that mGluR1/mGluR5-mediated activation of protein kinase C could enhance synaptic transmission by a number of mechanisms, including potentiation of current through N-type Ca^{2+} channel, disengagement of autoreceptor-mediated negative feedback (Schwartz et al., 1993), and facilitation of neurotransmitter release mechanism (Nicholls, 1993). This in combination with mGluR1/mGluR5-mediated mobilization of intracellular Ca^{2+} could lead to Ca^{2+} overload and subsequent activation of toxic events culminating in neuronal degeneration.

In addition to directly mediating neurotoxicity, in adult rats, mGluR activation by 1S,3R-ACPD can potentiate NMDA neurotoxicity (McDonald and Schoepp, 1992). This potentiation may be the result of the ability of mGluR activation to augment NMDA currents and responses (Aniksztejn et al., 1992; Harvey and Collingride, 1993; Kelso et al., 1992), and it may play a critical role in excitotoxicity. In this regard, it is intriguing to suggest that during excitotoxicity, activation of the NMDA receptor may be augmented to an abnormal level owing to potentiation via mGluR. Support for this notion could be derived from a recent observation that the mGluR antagonist L-AP3, when added during the insult, was found to confer protection against ischemia/hypoxia-induced injury (Opitz et al., 1993). Additionally, inhibition of mGluR1/mGluR5 effectors, i.e., protein kinase C (Felipo et al., 1993) and Ca^{2+} mobilization, can prevent glutamate toxicity (Frandsen and Schousboe, 1993).

6. Discussion

Available evidence suggests that mGluR activation has multiple effects on the survival of neurons. Activation of mGluRs can lead to neurotoxicity or neuroprotection. Considering the heterogeneity and the plasticity of the mGluR population, it is perhaps not surprising that different consequences of mGluR activation are observed. Although at present it is not possible to identify specifically mGluR subtypes responsible for mediating specific effects, available data can suggest that mGluR-induced neurotoxicity and the neuronal survival-promoting actions are mediated predominantly via an mGluR coupled to phospholipase C, and that the neuroprotective action may be mediated by mGluRs, negatively coupled to adenylyl cyclase and/or Ca^{2+} channels.

In acute neurological insult, mGluR activation may in some brain areas facilitate excitotoxicity by virtue of its ability to enhance neurotransmitter release and, at a postsynaptic level, potentiate NMDA receptor function. In regions where class II mGluR is functionally more dominant, mGluR activity may confer to that brain area or neuronal pathway resistance to excitotoxic damage. Finally, disruption or abnormal survival-promoting action of mGluR may play a role in chronic neurodegenerative diseases where the excitotoxic process develops more gradually.

References

Aniksztejn, L., Otani, S., and Ben-Ari, Y. (1992) Quisqualate metabotropic receptors modulate NMDA currents and facilitate induction of long-term potentiation through protein kinase C. *Eur. J. Neurosci.* **4**, 500–504.

Beal, M. F. (1933) Mechanisms of excitotoxicity in neurological diseases. *FASEB J.* **6**, 3338–3344.

Birrell, G. J., Gordon, M. P., Schwarz, R. D., and Marcoux, F. W. (1993) Metabotropic glutamate receptor mediated attenuation of NMDA-induced neuronal cell death in cerebrocortical cultures. *Functional Neurobiol.* **8**, 10,11.

Bruno, V. M. G., Copani, A., Battaglia, G., Marinozzi, M., Natalini, B., Pellicciari, R., . Kozikowski, A. P., Giffard, R., Choi, D. W., and Nicoletti, F. (1993) Effects of mGluR activation on different degenerative processes in cultured cells. *Functional Neurobiol.* **8**, 10,11.

Chiamulera, C., Albertini, P., Valerio, E., and Reggiani, A. (1992) Activation of metabotropic receptors has a neuroprotective effect in a rodent model of focal ischaemia *Eur. J. Pharmacol.* **216**, 335,336.

Choi, D. W. (1988) Glutamate neurotoxicity and diseases of the nervous system. *Neuron* **1**, 623–634.

Connor, J. A. and Tseng, H. Y. (1988) Measurement of intracellular Ca^{2+} in cerebellar Purkinje neurons in cultures: resting distribution and response to glutamate. *Brain Res. Bull.* **21**, 353–361.

Dubek, S. M., Bowen, W. D., and Bear, M. F. (1989) Postnatal changes in stimulated phosphoinositide turnover in rat neorcortical synaptoneurosomes. *Dev. Brain Res.* **47**, 123–128.

Evans, M. C., Griffiths, T., and Meldrum, B. S. (1983) Early light changes in the rat hippocampus following seizures induced by bicuculline or L-allylglycine. A light microscope study. *Neuropathol. Appl. Neurobiol.* **9**, 39–52.

Fagni, L., Bossu, J. L., and Bockaert, J. (1991) Activation of a large conductance Ca^{2+}-dependent K^+ channel by stimulation of glutamate phosphoinositide-coupled receptors in cultured cerebellar granule cells. *Eur. J. Neurosci.* **3**, 778–789.

Favaron, M., Manev, R. M., Candeo, P., Arban, R., Gabellini, N., Kozikowski, A. P., and Manev, H. (1993) Trans-azetidine-2,4-dicarboxylic acid activates neuronal metabotropic receptors. *NeuroReport* **4**, 967–970.

Felipo, V., Minana, M.-D., and Grisolia, S. (1993) Inhibitors of protein kinase prevent the toxicity of glutamate in primary neuronal cultures. *Brain Res.* **604**, 192–196.

Fix, A. S., Schoepp, D. D., Olney, J. W., Vestre, W. A., Griffey, K. I., Johnson, J. A., and Tizzano, J. P. (1993) Neonatal exposure to D,L-2-amino-3-phosphonpropionate (D,L-AP3) produces lesions in the eye and optic-nerve of adult-rats. *Dev. Brain Res.* **75**, 223–233.

Frandsen, A. and Schousboe, A. (1991) Dantrolene prevents glutamate neurotoxicity and Ca^{2+} release from intracellular stores. *J. Neurochem.* **56**, 1075–1078.

Frandsen A., and Schousboe, A. (1993) Excitatory amino acid-mediated cytotoxicity and calcium homeostasis in cultured neurons. *J. Neurochem.* **60**, 1202–1211.

Garthwaite, G. and Garthwaite, J. (1986) Neurotoxicity of excitatory amino acid receptor agonists in rat cerebellar slices: dependence on calcium concentration. *Neurosci. Lett.* **66,** 193–198.

Garthwaite, G. and Garthwaite, J. (1989a) Differential dependence on Ca^{2+} of *N*-methyl-D-aspartate and quisqualate neurotoxicity in young rat hippocampal slices. *Neurosci. Lett.* **97,** 316–322.

Garthwaite, G. and Garthwaite, J. (1989b) Quisqulate neurotoxicity: a delayed, CNQX-sensitive process triggered by a CNQX-insensitive mechanism in young rat hippocampal slices. *Neurosci. Lett.* **99,** 113–118.

Harvey, J., and Collingride, G. L (1993) Signal transduction pathways in the acute potentiation of NMDA responses by 1S,3R-ACPD in rat hippocampal slices. *Br. J. Pharmacol.* **109,** 1085–1090.

Hoffmann, M., Hartley, M., and Heinemann, S. (1991) Calcium permeability of KA-AMPA-gated glutamate receptor channels depends on subunit composition. *Science* **252,** 851–853.

Irving, A. J., Collingride, G. L., and Schofield, J. G. (1992) Interactions between Ca^{2+} mobilizing mechanisms in cultured rat cerebellar granule cells. *J. Physiol.* **456,** 667–680.

Judge, M. E., Sheardown, M. J., Jacobsen, P., and Honore, T. (1991) Protection against post-ischemic behavioral pathology by the alpha-amino-3-hydroxy-5-methyl-4-isoxazolepropionic acid (AMPA) antagonist 2,3-dihydroxy-6-nitro-7-sulfamoyl-benzo(f)quinoxaline (NBQX) in the gerbil. *Neurosci. Lett.* **133,** 291–294.

Kelso, S. R., Nelson, T. E., and Leonard, J. P. (1992) Protein kinase C-mediated enhancement of NMDA currents by metabotropic glutamate receptors in xenopus oocytes. *J. Physiol. (Lond.)* **449,** 705–718.

Koh, J.-K., Goldberg, M. P., Hartley, D. M., and Choi, D. W. (1990) Non-NMDA receptor-mediated neurotoxicity in cortical culture. *J. Neurosci.* **10,** 693–705.

Koh, J. Y., Palmer, E., and Cotman, C. W. (1991) Activation of the metabotropic glutamate receptor attenuates N-methyl-aspartate neurotoxicity in cortical cultures. *Proc. Natl. Acad. Sci. USA* **88,** 9431–9435.

Lipartiti, M., Fadda, E., Savoini, G., Siliprandi, R., Sautter, J., Arban, R., and Manev, H. (1993) In rats, the metabotropic glutamate receptors-triggered hippocampal neuronal damage is strain-dependent. *Life Sci.* **52,** PL85–90.

Mayer, M. L. and Miller, R. J. (1991) Excitatory amino acid receptors, second messengers and regulation of intracellular calcium in mammalian neurons. *Trends Pharm. Sci.* **11,** 36–42.

McCaslin, P. P. and Smith, T. G. (1988) Quisqualate, high calcium concentration and zero-chloride prevent kainate-induced toxicity of cerebellar granule cells. *Eur. J. Pharmacol.* **152,** 341–346.

McDonald, J. W. and Schoepp, D. D. (1992) The metabotropic excitatory amino acid receptor agonist 1S,3R-ACPD selectively potentiates *N*-methyl-D-aspartate-induced brain injury. *Eur. J. Pharmacol.* **215,** 353,354.

McDonald, J. W., Fix, A. S., Tizzano, J. P., and Schoepp, D. D. (1993) Seizures and brain injury in neonatal rats induced by 1S,3R-ACPD, a metabotropic glutamate receptor agonist. *J. Neurosci.* **13,** 4445–4455.

Meldrum, B. S. and Garthwaite, J. (1990) Excitatory amino acid neurotoxicity and neurodegenerative disease. *Trend Pharm. Sci.* **11,** 379–387.

Meyer, F. B. (1989) Calcium, neuronal hyperexcitability and ischemic injury. *Brain Res. Rev.* **14,** 227–243.

Mount, H. T. J., Dreyfus, C. F, and Black, I. B. (1993) Purkinje cell survival is differentially regulated by metabotropic and ionotropic excitatory amino acid receptors. *J. Neurosci.* **13,** 3173–3179.

Murphy, S. N. and Miller, R. J. (1988) A glutamate receptor regulates Ca^{2+} mobilization in hippocampal neurons. *Proc. Natl. Acad. Sci. USA* **85,** 8737–8741.

Nicholls, D. G. (1993) The glutamatergic nerve terminal. *Eur. J. Biochem.* **212,** 613–631.

Nicoletti, F., Iadarola, M. J., Wroblewski, J. T, and Costa, E. (1986) Excitatory amino acid recognition sites coupled with inositol phospholipid metabolism: Developmental changes and intereaction with alpha-1 adrenoceptors. *Proc. Natl. Acad. Sci. USA* **83,** 1931–1935.

Olney, J. W. (1978) Neurotoxicity of excitatory amino acids, in *Kainic Acid as a Tool in Neurobiology,* (McGeer, E. G., Olney, J. W., and McGeer, P. L., eds.) Raven, New York, pp. 1–15.

Olney, J. W., Price, M. T., Izumi, Y., and Romano, C. (1993) Neurotoxicity associated with either supression or excessive stimulation of mGluR function. *Lab. Invest.* **68,** 38,39.

Opitz, T., Hartmann, P., Richter, P., and Reymann, K. (1993) Metabotropic glutamate receptors are involved in hypoxic and ischemic injury of hippocampal CA1 neurons in vitro. *Functional Neurobiol.* **8,** 39,40.

Palmer, E., Nangel-Taylor, K., Krause, J. D., Roxas, A., and Cotman, C. W. (1990) Changes in excitatory amino acid modulation of phosphoinositide metabolism during development. *Dev. Brain Res.* **51,** 132–134.

Pizzi, M., Fallacara, C., Arrighi, V., Memo, M., and Spano, P. (1993) Attenuation of excitatory amino acid toxicity by metabotropic glutamate receptor agonists and aniracetam in primary cultures of cerebellar granule cells. *J. Neurochem.* **61,** 683–689.

Pizzi, M. Ribola, M., Valerio, A., Memo, M., and Spano, P. F. (1991) Various Ca^{2+} entry blockers prevent glutamate-induced neurotoxicity. *Eur. J. Pharmacol.* **109,** 169–173.

Prehn, J. H. M., Backhaub, C., Karkoutly, C. Nuglisch, J. Peruche, B., Roberg, C., and Krieglstein, J. (1991) Neuroprotective properties of 5-HT1a receptor agonists in rodent models of focal and global cerebral ischemia. *Eur. J. Pharmacol.* **203,** 213–222.

Randall, A. D., Wheeler, D. B., and Tsien, R. W. (1993) Modulation of Q-type Ca^{2+} channels and Q-type channel-mediated synaptic transmission by metabotropic and other G-protein linked receptors. *Functional Neurobiol.* **8,** 44,45.

Rothman, S. M. (1984) Synaptic release of excitatory amino acid neurotransmitter mediates anoxic neuronal death. *J. Neurochem.* **4,** 1884–1891.

Rudolphi, K. A., Schubert, P., Parkinson, F. E., and Fredholm, B. B. (1992) Neuroprotective role of adenosine in cerebral ischemia. *Trends Pharm. Sci.* **13,** 439–445.

Saccan, A. I. and Schoepp, D. D. (1992) Activation of hippocampal metabotropic excitatory amino acid receptors leads to seizures and neuronal damage. *Neurosci. Lett.* **139,** 77–82.

Sahara, Y. and Westbrook, G. L. (1993) Modulation of calcium currents by a metabotropic glutamate receptor involves fast and slow kinetic components in cultured hippocampal neurons. *J. Neurosci.* **13,** 3041–3050.

Schwartz, R. D., Birrell, G. J., and Marcoux, F. W. (1993) Involvement of metabotropic glutamate receptors in glutamate-induced neurotoxicity using rat cerebrocortical cultures. *Functional Neurobiol.* **8,** 50,51.

Siesjo, B. K. and Bengtsson, F. (1989) Calcium fluxes, calcium antagonists, and calcium-related pathology in brain ischemia, hypoglycemia, and spreading depression: a unifying hypothesis. *J. Cereb. Blood Flow Metab.* **9,** 127–140.

Siliprandi. R., Lipartiti, M., Fadda, E., Sautter, J., and Manev, H. (1992) Activation of the glutamate metabotropic receptor protects retina against N-methyl-D-aspartate toxicity 1S,3R-ACPD ((1S,3R)-1-aminocyclopentane-1,3-dicarboxylic acid) metabotropic glutamate receptors excitotoxicity. *Eur. J. Pharmacol.* **219,** 173,174.

Silverstein, F. S., Chen, R, and Johnston, M. V. (1986) The glutamate analogue quisqualic acid is neurotoxic in striatum and hippocampus of immature rat brain. *Neurosci. Lett.* **71,** 13–18.

Simon, R. P., Griffiths, T., Evans, M. C., Swan, J. H., and Meldrum, B. S. (1984) Calcium overload in selectively vulnerable neurons of the hippocampus during and after ischemia: an electron microscopy study in the rat. *J. Cereb. Blood Flow Metab.* **4,** 350–361.

Sloviter, R. S. and Dempster, D. W. (1985) Epileptic brain damage is replicated qualitatively in the rat hippocampus by central injection of glutamate or aspartate but not by GABA or acetylcholine. *Brain Res. Bull.* **15,** 39–60.

Tymianski, M., Wallace, M. C., Spigelman, I., Uno, M., Carlen, P. L., Tator, C. H., and Charlton, M. P. (1993) Cell-permeant calcium chelators reduce early excitotoxic and ischemic neuronal injury in vitro and in vivo. *Neuron* **11,** 221–235.

Valentino K., Newcomb, R., and Gadbois, T. (1993) A selective N-type calcium channel antagonist protects against neuronal loss after global ischemia. *Proc. Natl. Acad. Sci. USA* **90,** 7894–7897.

Zinkand, W. C., DeFeo, P. A., Thompson, C., Hargrove, H., Salama, A. I., and Patel, J. (1992) Quisqualate neurotoxicity in rat cortical cultures: pharmacology and mechanisms. *Eur. J. Pharm.* **212,** 129–136.

CHAPTER 10

Plasticity of Metabotropic Glutamate Receptors in Physiological and Pathological Conditions

*F. Nicoletti, E. Aronica, G. Battaglia,
V. Bruno, G. Casabona, M. V. Catania,
A. Copani, A. A. Genazzani,
M. R. L'Episcopo, and D. F. Condorelli*

1. Metabotropic Glutamate Receptors (mGluRs) in Brain Slices

Information on native mGluRs has been provided by studies performed in brain slices, synaptoneurosomes, and primary cultures of neurons or astrocytes. In brain slices, mGluRs are either coupled to phosphatidylinositol (PI) turnover (mGluR$_{PI}$) (Nicoletti et al., 1986a; Schoepp and Johnson, 1988) or negatively linked to adenylyl cyclase activity (mGluR$_{\downarrow cAMP}$) (Cartmell et al., 1992; Schoepp et al., 1992; Genazzani et al., 1993). Recently, two independent groups have shown that mGluR agonists activate phospholipase D (Boss and Conn, 1993; Holler et al., 1993), a process that generates large

The Metabotropic Glutamate Receptors Eds.: P. J. Conn and J. Patel
© 1994 Humana Press Inc., Totowa, NJ

amounts of diacylglycerol as a result of phosphatidylcholine hydrolysis and phosphatidic acid formation. The selective mGluR agonist (1S,3R)-1-aminocyclopentane-1,3-dicarboxylic acid (1S,3R-ACPD) potentiates cAMP responses to agonists of G_s-coupled receptors, and increases basal cAMP formation (Winder and Conn, 1992), by potentiating the response to endogenous adenosine acting at A2 purinergic receptors (Winder and Conn, 1993; Cartmell et al., 1993; *see* Chapter 3).

The pharmacological profile of mGluR$_{PI}$ in brain slices is substantially different from that reported in cells transfected with mGluR1 or mGluR5, the two cloned subtypes coupled to PI turnover (reviewed in Nakanishi, 1992). In brain slices, (2S,1'S,2'S)-2-(carboxycyclopropyl)glycine (L-CCG-I), 1S,3R-ACPD, and ibotenate are more efficacious than quisqualate in stimulating [^{3}H]inositol monophosphate (IP) formation, whereas cysteic acid, L-cysteinylsulfinate, L-homocysteinsulfinate, D-aspartate, the Guam toxin β-methylamino-L-alanine, L-aspartate, and L-glutamate all exhibit low intrinsic efficacy (Nicoletti et al., 1986a; Schoepp et al., 1990; Nakagawa et al., 1990; Copani et al., 1990, 1991; Porter and Roberts, 1993; Littman et al., 1993). L-2-Amino-3-phosphonopropionate (AP3), L-2-amino-4-phosphonobutanoate (AP4), L-serine-*O*-phosphate, and L-aspartate-β-hydroxamate act as noncompetitive antagonists of mGluR$_{PI}$ in brain slices (Nicoletti et al., 1986a,b; Schoepp and Johnson, 1988, 1989; Littmann et al., 1992), an effect that is hardly detectable in transfected cells. (RS)-α-methyl-4-carboxyphenylglycine (MCPG) and other carboxyphenylglycine derivatives are so far the only available competitive antagonists of mGluR$_{PI}$ in brain slices (Eaton et al., 1993).

The pharmacological profile of mGluR$_{PIS}$ is different from that exhibited by mGluR$_{\downarrow cAMPS}$. The most striking difference concerns the actions of AP3, AP4, and L-SOP, which inhibit mGluR$_{PI}$, but are as efficacious as 1S,3R-ACPD in reducing forskolin-stimulated cAMP formation (i.e., in activating mGluR$_{\downarrow cAMPS}$) (Casabona et al., 1992; Schoepp and Johnson, 1993; Genazzani et al., 1993). 1R,3S-ACPD, which is virtually inactive on mGluR$_{PIS}$ (Schoepp et al., 1991), is even more efficacious (although less potent) than 1S,3R-ACPD in reducing forskolin-stimulated cAMP formation (Schoepp et al., 1992).

2. DCG-IV as a Potential Positive Modulator of mGluR$_{PI}$s in Brain Slices

A series of carboxycyclopropylglycine derivatives have been introduced as selective or mixed mGluR ligands. Among these, L-CCG-I acts as a mixed agonist of mGluR1 and mGluR 2 subtypes in transfected cells (Nakanishi, 1992). (2S,1'R,2'R,3'R)-2-(2,3-Dicarboxycyclopropyl)glycine (DCG-IV) is a novel mGluR ligand capable of reducing synaptic transmission in the spinal cord (Ohfune et al., 1993; Ishida et al., 1993). In transfected cells, DCG-IV behaves as a potent and selective agonist of the mGluR2/3 subtypes, with very low activity on mGluR4 and no activity on mGluR1 (Hayashi et al., 1993). In brain slices, DCG-IV is by far the most potent mGluR agonist in inhibiting forskolin-stimulated cAMP formation (Genazzani et al., 1993; Fig. 1). We have tested the effect of DCG-IV on the functional expression of mGluR$_{PI}$s in hippocampal slices. DCG-IV does not affect the basal PI turnover, but substantially enhances both the potency and efficacy of quisqualate in stimulating [^{3}H]IP formation (Fig. 2). DCG-IV also increases the potency—but not the efficacy—of ibotenate or 1S,3R-ACPD and is virtually inactive on carbamylcholine-stimulated PI turnover (*see* Nicoletti et al., 1993). The modulatory action of DCG-IV is already visible at concentrations as low as 10 nM, and is maximal between 1 and 10 μM (Nicoletti et al., 1993). Inhibition of cAMP formation does not contribute to the enhancement of quisqualate-stimulated PI turnover by DCG-IV. The action of DCG-IV is neither reversed by forskolin nor mimicked by baclophen or somatostatin, i.e., by agonists of receptors that are negatively linked to adenylyl cyclase (Table 1).

The large NH$_2$-terminal domain of mGluR subtypes coupled to PI turnover suggests the existence of a modulatory site that recognizes DCG-IV as a ligand. However, the identity of the receptor subtype sensitive to the action of DCG-IV remains to be established. DCG-IV does not enhance quisqualate-stimulated PI turnover in primary cultures of cerebellar granule cells, which predominantly express mGluR1 receptors (Bessho et al., 1992; Aronica et al., 1993), or in mixed cultures of cortical neurons and astrocytes (cortical neu-

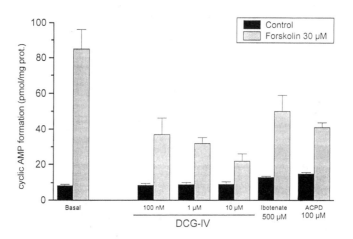

Fig. 1. Inhibition of forskolin-stimulated cAMP formation by the novel mGluR ligand DCG-IV in adult rat hippocampal slices. cAMP formation was determined in slices incubated for 30 min in the presence of forskolin and mGluR ligands, as described previously (Genazzani et al., 1993). MK-801 (10 μM) was added to the incubation buffer to prevent the stimulation of NMDA receptors by 10 μM DCG (*see* Ishida et al., 1993). Values are means ± SEM of four to six determinations.

rons express both mGluR1 and mGluR5—Masu et al., 1991; Abe et al., 1992). Radioligand binding studies will help elucidate the nature of the interaction between DCG-IV and mGluR$_{PI}$ in hippocampal slices.

3. Developmental Pattern of Expression of mGluRs in Brain Slices

mGluR$_{PIS}$ undergo plastic changes during postnatal development. Stimulation of [^{3}H]IP formation by all known mGluR agonists is very robust during the earlier stages of postnatal life in slices prepared from various brain regions, including hippocampus, cerebral cortex, corpus striatum, hypothalamus, cerebellum, and olfactory bulb (Nicoletti et al., 1986a; Schoepp and Johnson, 1989; Schoepp and Hillman, 1990; Sortino et al., 1991; Vecil et al., 1992; Condorelli

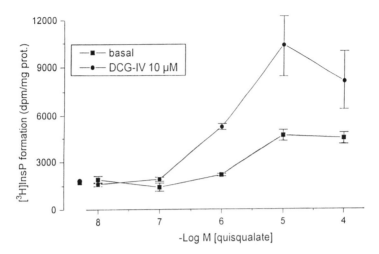

Fig. 2. Concentration-dependent stimulation of PI turnover by quisqualate in adult rat hippocampal slices incubated in the absence or presence of 10 μM DCG-IV. For measurement of PI turnover, *see* legend to Table 1. MK-801 (10 μM) was added to prevent the stimulation of NMDA receptors by DCG-IV. Values are means ± SEM of six determinations from two individual experiments.

et al., 1992). The PI response to mGluR agonists increases from 1 to 7–10 d after birth, and then declines progressively across maturation to reach adult values after the 24th d of postnatal life. In contrast, stimulation of PI turnover by carbamylcholine does not change substantially, whereas stimulation by norepinephrine increases with age (Nicoletti et al., 1986a). The sensitivity of mGluR$_{PIS}$ to noncompetitive antagonists also varies with age. L-AP4 and L-SOP act as noncompetitive antagonists of mGluR$_{PIS}$ in adult slices, whereas they are virtually inactive in slices prepared from 7–15-d-old rats. The lower PI response to glutamate in the adulthood cannot be simply ascribed to a greater efficiency of the uptake system for excitatory amino acids, since the efficacy, rather than the potency, of glutamate decreases with age. In addition, postnatal maturation is accompanied by a 7–10-fold reduction in the PI response to ibotenate, which is not a substrate for the uptake carrier. In adults, all the mGluR$_{PI}$ agonists

Table 1
Stimulation of PI Turnover by Quisqualate
in Adult Rat Hippocampal Slices Treated with Agonists
of Receptors Negatively Linked to Adenylyl Cyclase

	[^{3}H]IP formation, dpm/mg prot.	
	Control	Quisqualate, 100 μM
Basal	1800 ± 110	3300 ± 650
DCG-IV, 10 μM	1900 ± 80	8300 ± 1100
Somatostatin, 10 μM	1700 ± 220	3700 ± 150
Baclophen, 100 μM	2000 ± 180	3900 ± 500

Values are means ± SEM of six determinations. [^{3}H]IP (inositol monophosphate) formation was determined in slices prelabeled with myo-d-[^{3}H]-
inositol and incubated for 60 min in the absence of presence of quisqualate,
as described previously (Nicoletti et al., 1986a).

that have low efficacy, such as quisqualate and glutamate, as well as
L-AP4 and L-SOP, inhibit norepinephrine-stimulated PI turnover
without affecting [^{3}H]prazosin binding to α_1-adrenergic receptors.
The molecular mechanism responsible for this effect is unclear at
present. The possibility of an interactive coupling among mGluR$_{PI}$,
α_1-adrenergic receptors, and phospholipase C has been suggested.

Dudek and Bear (1989a,b) have shown that, in the kitten visual
cortex, the functional activity of mGluR$_{PI}$s is greater between 3 and
5 wk of postnatal life, a time that corresponds to the "critical period"
of synaptic modification in response to monocular deprivation.
Developmental changes in mGluR$_{PI}$ activity are abolished if kittens
are reared in complete darkness, suggesting that changes in mGluR
activity are in relation to activity-dependent modification in synaptic efficacy.

Age-related changes in mGluR activity have been also shown in
other models, including synaptoneurosomes (Guiramand et al., 1989)
and neurons grown in primary cultures (Weiss et al., 1988; Lin et
al., 1991; Aronica et al., 1993a; *see* Section 4.). As opposed to
mGluR$_{PI}$, the activity of mGluR$_{\downarrow cAMP}$s increases with age (Casabona

et al., 1992). Thus, the ability of 1S,3R-ACPD to reduce forskolin-stimulated cAMP formation in hippocampal or hypothalamic slices is virtually absent at 7 or 15 d after birth, whereas it is present in adults (Fig. 3).

Three fundamental questions need to be addressed:

1. Which receptor subtype is responsible for the typical developmental pattern of mGluR$_{PI}$ activity in brain slices;
2. Which molecular mechanism is involved in the progressive decline of receptor activity across maturation; and
3. What is the functional significance of the developmental changes in mGluR$_{PI}$ activity.

There is no correlation between the PI response to 1S,3R-ACPD in slices from discrete brain regions and steady-state levels of mGluR1α mRNA (Condorelli et al., 1992). Thus, responses to 1S,3R-ACPD are greater in 7-d-old slices from hippocampus or hypothalamus, whereas mGluR1α mRNA levels are higher in adult cerebellum and olfactory bulb. mGluR1α mRNA levels are stable across maturation in the hypothalamus, cerebral cortex, and olfactory bulb, whereas they increase with age in the hippocampus and cerebellum (Condorelli et al., 1992). This suggests that, in brain slices, the reduced expression of mGluR$_{PI}$ activity during postnatal development is not the result of a decreased rate of transcription of the gene encoding the mGluR1α. A better correlation exists between responses to agonists and mGluR5 mRNA levels, which are highly expressed at birth and decrease during the first 3 wk of postnatal life (Catania et al., 1993). Studies on the molecular mechanisms and functional significance of the developmental changes in mGluR$_{PI}$ activity are difficult in brain slices because of the heterogeneity of cell and receptor population (*see* Schoepp and Hillmann, 1990; Vecil et al., 1992; Condorelli et al., 1992). For this reason, we have addressed questions 2 and 3 in primary cultures of cerebellar neurons, which are highly enriched (>90%) in granule cells and express predominantly—although not exclusively—the mGluR1α subtype coupled to PI turnover (Favaron et al., 1992; Bessho et al., 1993; Aronica et al., 1993a).

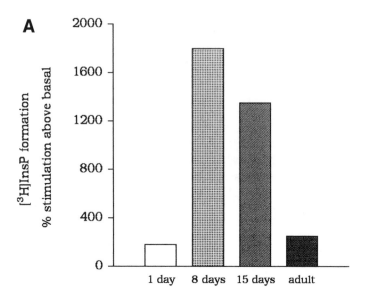

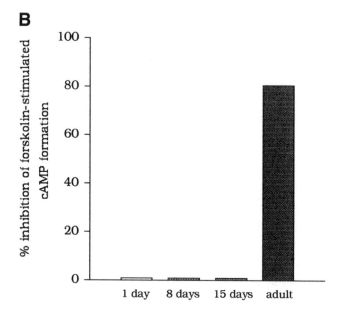

Fig. 3. Developmental changes in the ability of 1S,3R-ACPD to stimulate PI turnover (**A**) or to inhibit forskolin-stimulated cAMP formation (**B**) in rat hippocampal slices. On the *x* axis: days after birth. Basal [³H]IP formation was reduced with age (*see* Nicoletti et al., 1986c and Condorelli et al., 1992).

4. Developmental Pattern of Expression of mGluR$_{PI}$ in Primary Cultures of Cerebellar Neurons

In cultured cerebellar granule cells, quisqualate and glutamate stimulate PI turnover with a greater potency and efficacy than 1S,3R-ACPD (Aronica et al., 1993b). This pharmacological profile is similar to that reported in cells transfected with mGluR1α cDNA (Aramori and Nakanishi, 1992). In addition, Northern analysis reveals the presence of substantial amounts of mGluR1α mRNA levels, although mGluR5 mRNA levels are detectable using a competitive PCR-derived assay (Wroblewski et al., 1993). In cultures grown under "standard" conditions (i.e., in medium containing 10% fetal calf serum and 25 mM K$^+$), stimulation of PI turnover by quisqualate, glutamate, or 1S,3R-ACPD is developmentally regulated and peaks after 4 d of maturation in vitro (DIV) before declining at later stages of maturation. This pattern is shared by mGluR1α mRNA levels, which are higher at 3 than at 5, 7, or 9 DIV (Aronica et al., 1993a). Treatment of cultures with the enzyme alanine aminotransferase, which depletes extracellular glutamate, prevents the developmental decline in mGluR activity, suggesting that the receptor is desensitized by the endogenous glutamate. Interestingly, the process of homologous desensitization itself is developmentally regulated and begins to occur only after 4 DIV (Aronica et al., 1993c). Thus, pre-exposure of cultures to 1–10 µM glutamate (a concentration found "physiologically" in the medium) reduces the PI response to a successive challenge with glutamate or quisqualate in cultures at 7 DIV, but not in cultures at 4 DIV. The negative feedback of protein kinase C on mGluR activity, which is the basis for homologous desensitization, is also developmentally regulated (Aronica et al., 1993c). It is therefore possible that the development of homologous desensitization sets the clock, which regulates mGluR activity during neuronal maturation. Such a finely tuned mechanism of regulation suggests that mGluR$_{PI}$s regulate the transition among proliferation, differentiation, and "programmed cell death" (apoptosis), which can be paradoxically viewed as a specific differentiated phenotype (Wyllie et al., 1980). During the time of maximal mGluR activity (i.e., within the first 4 DIV), cultured cerebellar granule cells survive independently of the K^+ concentration in the medium. Afterward, chronic depolarization (provided by 25 mM K^+) is necessary for optimal growth and survival (Gallo et al., 1987; Balazs et al., 1988). When grown in 5 or 10 mM K^+, cells undergo degeneration within 2–3 d. We have shown that cultures grown in "low K^+" degenerate through an apoptotic pathway, characterized by chromatin fragmentation and condensation, a process that can be visualized by appropriate fluorescent nuclear staining. "Low-K^+"-induced apoptosis is prevented by the protein synthesis inhibitor cycloheximide, suggesting that this process falls into the category of programmed cell death. Interestingly, pharmacological activation of mGluRs after its peak of activity (i.e., from 5 to 7 DIV) protects cultured cerebellar granule cells against apoptotic death (Fig. 4 A,B). This effect is prevented by MCPG, but not by the NMDA receptor antagonist, MK-801 (Table 2). We suggest that a high activity of mGluRs during the earlier stages of development is necessary to protect neurons against apoptosis. Apoptosis would occur when receptor activity declines, unless neurons are properly innervated or their survival is supported by specific trophic factors.

5. Plastic Changes in mGluR$_{PI}$ Activity After Denervation

Stimulation of PI turnover by mGluR agonists is substantially enhanced in hippocampal slices prepared at different times after intrahippocampal infusion of kainate or colchicine (Nicoletti et al., 1987). Intrahippocampal injection of kainate preferentially kills pyramidal neurons in the CA3 region, which project to the Sommer's sector (CA1 regions) via the Schaffer collateral. The increase in the sensitivity of mGluR$_{PI}$s is already maximal 24 hours after kainate infusion—a time that precedes the occurrence of reactive gliosis—and lasts for at least 7 d. This effect is accompanied by an increased PI response to norepinephrine, whereas the activity of the muscarinic receptor agonist carbamylcholine is not affected by kainate infusion.

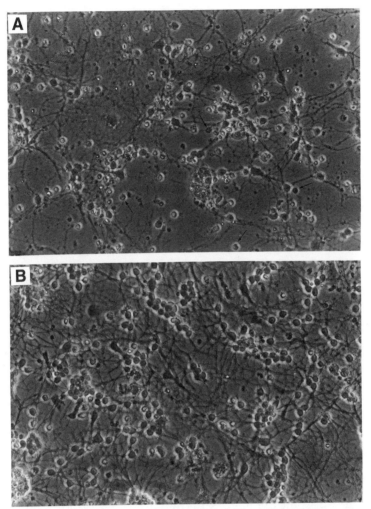

Fig. 4. **(A)** Phase-contrast micrographs of cultured cerebellar neurons (7 DIV) grown in medium containing 10 mM K$^+$; **(B)** same as in (A) with 1S,3R-ACPD (200 μM) added to the culture only once at 2 DIV. Arrows show examples of apoptotic neurons.

A similar increase in mGluR$_{PI}$ activity is observed after intrahippo-campal injection of colchicine, which selectively destroys granule cells of the dentate gyrus, as well as in the striatum after mechanical ablation of the frontal cortex (i.e., after lesioning the cortico-striatal

Table 2
Activation of mGluRs Prevents Apoptosis by Trophic
Deprivation in Primary Cultures of Cerebellar Granule Cells

	Apoptotic neurons, % of K10 values
Cultures Grown in	
25 mM K$^+$	12 ± 1
10 mM K$^+$ (K10)	100 ± 13
K10 + 1S,3R-ACPD (200 μM)	25 ± 7
K10 + MCPG (1 mM)	74 ± 21
K10 + ACPD + MCPG	103 ± 12

ACPD or MCPG were added after 2 d of maturation in vitro (DIV), and apoptosis was determined at 7 DIV. Chromatin fragmentation and/or condensation was revealed by fluorescent nuclear staining with the dye Hoechst 33258.

afferent pathways) (Nicoletti et al., 1987). Taken collectively, these results suggest that denervation restores a pattern of mGluR$_{PI}$ activity that is typical of the early postnatal development, supporting a critical role for this receptor in the neuronal (and/or glial) response to injury. Once again, the increased functional activity of the receptor does not correlate with the steady-state levels of mGluR1α mRNA, which are substantially reduced 7 d after kainate infusion in the hippocampus (Fig. 5a,b).

6. Plastic Changes of mGluR$_{PI}$ Activity During Acute or Chronic Neurodegenerative Diseases

The substantial changes in mGluR$_{PI}$ activity observed during maturation or in response to denervation suggests that this specific receptor may participate in the transsynaptic or intracellular mechanisms that accompany the evolution of acute and chronic neurodegenerative disorders, in which increases in intracellular Ca^{2+} and activation of protein kinase C—i.e., the intracellular consequences of mGluR$_{PI}$ activation—have been widely implicated both in the pathophysiology of neuronal damage and in neuronal repair. Chen et al. have shown in 1988 that quisqualate-stimulated PI turnover is

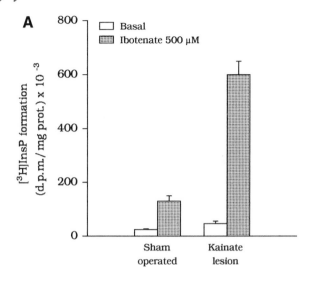

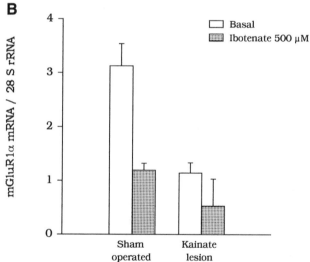

Fig. 5. Stimulation of PI turnover by ibotenate in hippocampal slices (**A**) or hippocampal mGluR1α mRNA levels (**B**) 15 d after local infusion of kainate. Kainate was injected unilaterally in the hippocampal formation (0.5 μg/0.5 μL/2 min), as described previously (Nicoletti et al., 1987). Northern analysis of mGluR1α mRNA was performed using a full-length probe kindly provided by Dr. M Masu and S. Nakanishi (Kyoto University, Japan), as described by Condorelli et al. (1992). Results were normalized by the amount of 28S rRNA.

enhanced in brain slices from newborn rats that underwent a hypoxic-ischemic insult. We have confirmed these results in adult rats subjected to transient global ischemia induced by permanent occlusion of the vertebral arteries followed by transient occlusion of the carotid arteries (Pulsinelli's model), an experimental model of cardiac arrest in humans. Under our experimental conditions, a 30-min period of global ischemia led to neuronal damage of the CA1 and CA4 regions of the hippocampus, and occasional damage of pyramidal neurons in the cerebral cortex, whereas neurons in the striatum were spared. Stimulation of PI turnover by mGluR agonists and norepinephrine was substantially enhanced in slices prepared 24 h or 7 d (but not 1 h) after postischemic recirculation, suggesting that ischemia leads to a slowly developing increase in the functional activity of mGluR$_{PI}$ (Seren et al., 1989). This effect is restricted to vulnerable regions, i.e., to the hippocampus and cerebral cortex. Using the same experimental model, Iversen et al. (1993) have found that the steady-state levels of mGluR1 and mGluR5 are reduced after ischemia in the CA3 and CA1. Studies with specific antibodies are needed to determine which component of the receptor complex is involved in the plastic changes of mGluR$_{PI}$ activity after ischemia.

An increase in the sensitivity of mGluR$_{PI}$s to agonists is also observed in slices exposed to various periods of anoxia in vitro (Ninomiya et al., 1990), supporting the view that mGluR$_{PI}$s play an important role in the reaction of neurons and/or astrocytes to the hypoxic/ischemic insult. Activation of mGluR$_{PI}$s by the released glutamate during reperfusion of ischemic tissue may contribute to the "maturation" of the ischemic damage in combination with AMPA receptors, which may acquire the ability to increase Ca^{2+} influx owing to a reduced synthesis of the GluR2 subunit (Pellegrini-Giampietro et al., 1992). However, the evidence that 1S,3R-ACPD protects neurons against the ischemic damage suggests that mGluR$_{PI}$s may have a protective role or may therefore contribute to the reparative phenomena, which are promoted by postischemic recirculation.

A potential role for mGluR$_{PI}$s in chronic neurodegenerative disorders stems from the evidence that β-N-methylamino-L-alanine (BMAA), a nonprotein amino acid present in the seed of the false sago palm, *Cycas circinalis*, stimulates PI turnover in brain slices and neuronal cultures (Copani et al., 1990, 1991). BMAA-induced PI turnover is developmentally regulated, is not additive with that of ACPD, and is antagonized by L-AP3, suggesting that BMAA acts as an mGluR$_{PI}$ agonist (Table 3). It is consistent with this hypothesis that BMAA displaces specifically bound [^{3}H]glutamate to metabotropic receptor sites with an apparent K_i value of about 20 μM (Cha et al., 1990). Interaction between BMAA and mGluRs requires the presence of physiological concentrations of bicarbonate ions, which react with the β-amino group of the toxin to produce a carbamate adduct (*see* Table 3). BMAA has been implicated in the pathophysiology of amyotrophic lateral sclerosis-parkinsonism-dementia complex (ALS-PD) among the Chamorro population of the western Pacific islands of Guam and Rota, and induces neurodegenerative phenomena when administered orally to Macaques or added acutely to explants of mouse motor cortex (Spencer et al., 1987).

Changes in [^{3}H]glutamate binding to metabotropic receptors (defined as the binding left in the presence of 100 μM NMDA, 100 μM AMPA, 100 mM KSCN, and 2.5 mM CaCl$_2$) are found in autoptic samples from hippocampal regions of patients suffering Alzheimer's disease. In particular, mGluR binding was reduced in the subiculum and CA1, but not in the parahippocampal gyrus, although AMPA and kainate binding were reduced in the latter region (Dewr et al., 1991). Further studies on mGluR binding have led to differentiation of two populations of sites, named met-1 and 2, characterized by high and low affinity to quisqualate, respectively. Only binding to the met-2 site was reduced in the insular cortex of patients with Alzheimer's disease or combined Alzheimer's/Parkinson's disease, as well as in the substantia nigra of parkinsonian patients, whereas the met-1 binding was unaltered in all the regions examined (Young et al., 1993; *see* Chapter 4). It is therefore possible that autoptic changes in metabotropic binding in chronic neurodegenerative disorders mainly involve the subtypes negatively linked to adenylyl cyclase, which all exhibit a greater affinity for ACPD than for quisqualate.

Recently, Schoepp and collaborators have shown that intra-hippocampal injection of 1S,3R-ACPD induces limbic seizures and

Table 3
Stimulation of PI Turnover by the Guan Toxin BMAA
in Hippocampal Slices from 8-Day-Old Rats[a]

| | | [³H]IP formation, dpm/mg prot. × 10⁻² | | |
	Control	AP3 1 mM	D-AP5 1 mM	CNQX 250 μM
Basal	16 ± 0.3	40 ± 5	20 ± 5	11 ± 0.8
BMAA, 1 mM	160 ± 27	90 ± 15	180 ± 10	170 ± 38
t-ACPD, 100 μM	27 ± 6	110 ± 20	230 ± 30	240 ± 45
	25 mM HCO₃		1 mM HCO₃	
Basal	20 ± 2.7		18 ± 0.9	
BMAA, 1 mM	390 ± 36		80 ± 12	

[a]Values are means ± SEM of 6–9 determinations.

neuronal damage that was indirectly mediated by NMDA receptors (Sacaan and Schoepp, 1992). In addition, intrathalamic injection of 1S,3R-ACPD induces limbic seizures that are attenuated by L-AP3 or by dantrolene, an inhibitor of intracellular Ca^{2+} mobilization (Tizzano et al., 1993). The action of 1S,3R-ACPD is stereoselective and is insensitive to D-AP3, as well as to the ionotropic receptor antagonists MK-801 or GYKI 52466. L-AP3 is also able to inhibit audiogenic seizures when injected icv to DBA/2 mice (Klitgaard and Jackson, 1993). Taken collectively, these results suggest that activation of mGluR$_{PIS}$ may contribute to the pathophysiology of epileptic seizures. A role for mGluRs in the genesis of neuronal hyperexcitability had been suggested using the kindling model of epilepsy. Kindling refers to a condition resulting from repeated administrations of an initially subconvulsant stimulus that cause a progressive intensification of the response culminating in overt motor seizures. Stimulation of PI turnover by mGluR agonists is enhanced in hippocampal slices from animals that underwent electrical kindling of the hippocampus or amygdala (Iadarola et al., 1986; Akijama et al., 1987). The increased sensitivity of mGluR$_{PIS}$ begins to occur after stage 3 of Racine's scale, when electrical discharge and seizure activity spread and tend to be generalized. It is therefore possible that activation of mGluR$_{PI}$ facilitates the transmission of neuronal hyperactivity from a specific focus to the surrounding areas.

7. Plastic Changes of mGluRs During Learning and LTP Formation

Agonist-stimulated PI turnover has been studied in hippocampal slices prepared from animals trained in a eight-arm radial maze to perform a specific task, i.e., the search for a reward by entering new arms using extramaze spatial cues as reference stimuli. The activity of mGluR$_{PI}$s was greater in hippocampal slices prepared from trained than from untrained animals (Nicoletti et al., 1988). It was unclear, however, whether the learning process led to an increased sensitivity of mGluR$_{PI}$s or whether a higher mGluR$_{PI}$ activity was a basic feature of good learners. To address this question, we have studied the plastic changes of mGluR$_{PI}$ during the formation of "long-term potentiation" (LTP), an established electrophysiological model of associative learning. LTP has been induced either in the pyramidal cells of the CA1 region as a result of tetanic stimulation of the Schaffer collaterals in in vitro hippocampal slices, or in the granule cells of the dentate gyrus, as a result of tetanic stimulation of the perforant pathway in freely moving animals (Aronica et al., 1991). Stimulation of PI turnover by the mGluR agonists ibotenate and ACPD, but not by carbamylcholine, is enhanced in hippocampal slices at 5 or 12 h (but not at earlier times) after the induction of LTP, suggesting that LTP formation is accompanied by a slowly developing increase in the sensitivity of mGluR$_{PI}$s. This effect is abolished when tetanic stimulation is performed in the presence of NMDA receptor antagonists, which prevent the induction of LTP (Aronica et al., 1991). The late increase in mGluR$_{PI}$ activity during LTP formation is not associated with an increased [^{3}H]glutamate binding to metabotropic receptor sites with changes in the levels of mGluR1α mRNA (not shown), suggesting that changes in the coupling between recognition sites and phospholipase C are involved in the plastic changes of mGluR$_{PI}$s. Hyperactivity of mGluR$_{PI}$s during the late stages of LTP formation may be relevant for the changes in the cytoskeleton dynamics that accompany the consolidation of the memory trace and characterize the transition from short-term into long-term memory. Phosphatidylinositol-4,5-bis-phosphate, the substrate of mGluR-activated phospholipase C, tightly interacts with some of the major actin binding proteins, such as profilin and gelsolin, which control actin polymerization, severing of actin filaments, network organization, and actin filament–membrane interaction (Forscher, 1989). In addition, protein kinase C regulates the cycling of the actin binding protein MARCKS (myristoilated, alanine-rich C kinase substrate) between the plasma membrane and the cytosol, thus contributing to the remodeling of the cytoskeleton (see Thelen et al., 1991). Activation of mGluRs also occurs during the induction stage of LTP and is likely to play an important role in induction of LTP as discussed in detail in Chapter 7.

In conclusion, mGluR$_{PI}$s appear to have a pivotal role in the transsynaptic mechanisms that enable the induction, expression, and maintenance of LTP, and this suggests a new strategy in the therapy of learning and memory deficits during aging. Aniracetam, one of the nootropic drugs used as memory and cognition enhancers, has been found to enhance the functional expression of mGluR$_{PI}$s in cultured neurons (Pizzi et al., 1993).

8. Plasticity of mGluRs Coupled to PI Turnover in Astrocytes

mGluR$_{PI}$s are also expressed in cultured astrocytes, where their activation by quisqualate, glutamate, ibotenate, or ACPD leads to inositol phosphate formation and oscillatory increases in intracellular Ca^{2+} (Pearce et al., 1987; Milani et al., 1988; Condorelli et al., 1989a; Nicoletti et al., 1990; Jensen and Chiu, 1990). Quisqualate can also translocate and activate protein kinase C, as suggested by the increase in phorbol ester binding in intact cells. The identity of the mGluR (i.e., the specific subtype) that mediates this effect is unknown at present. mGluR agonists substantially reduce [^{3}H]thymidine incorporation and astrocyte proliferation in all the conditions examined, i.e., in astrocytes grown in the presence of serum, as well as in astrocytes deprived of serum for 24 h (i.e., arrested in G_0) and then stimulated to proliferate by addition of epidermal growth factor (EGF) or phorbol esters (Condorelli et al., 1989a; Nicoletti et al., 1990). In addition, both quisqualate and 1S,3R-ACPD increase glutamine synthetase activity, a glial-specific enzyme that contributes to the regulation of glutamate concentrations in the CNS (Nicoletti et al., 1990; Miller et al., 1991). The effect of 1S,3R-ACPD on glutamine synthetase activity is concentration-dependent and stereoselective, and is blocked by the protein synthesis inhibitor cycloheximide, suggesting that activation of mGluRs leads to gene expression and synthesis of new proteins in astrocytes (Miller et al., 1991). It is consistent with this hypothesis that both quisqualate and ACPD induce a rapid and transient increase in the expression of the primary response genes, c-*fos*, *fos*-B, *jun*B, and zif/268 (Condorelli et al., 1989b; Fig. 6), the products of which will influence the expression of secondary response genes, thus converting short-term extracellular stimuli into long-lasting changes in astrocyte morphology and function. All these effects emphasize the importance of neuronglia interaction and suggest that the glutamate released from neurons may contribute to regulation of the rate of proliferation, as well as the phenotypical and functional specialization of surrounding glial cells.

The importance of mGluR$_{PIS}$ in astrocytes is strengthened by the evidence that these receptors undergo plastic modifications in relation to the growing conditions, as well as to the rate of cell proliferation. Twelve-hour exposure of quiescent astrocytes to the proliferative agent, EGF, increased the potency of glutamate in stimulating PI turnover, an effect that is not observed at short times after EGF addition (Nicoletti et al., 1990). The plasticity of mGluRs in astrocytes has been confirmed by the elegant experiments by Miller et al. (1993), who measured the ability of ACPD to stimulate PI turnover in astrocytes cultured in a serum-free defined medium, as compared to astrocytes cultured in conventional serum-containing medium. The two conditions have profound influences on the morphology of astrocytes, which are flat and polygonal when grown in serum-containing medium, but stellate and highly branched when grown in

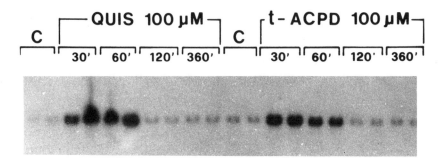

Fig. 6. mGluR agonists and kainate induce a rapid and transient increase in the expression of the protooncogene, zif/268 in cultured astrocytes. Astrocyte cultures were prepared and zif/268 mRNA levels measured as described by Condorelli et al. (1993).

chemically defined medium. Stimulation of PI turnover by ACPD is much greater (40 fold vs threefold) in cultures grown in chemically defined medium, whereas inhibition of forskolin-stimulated cAMP formation or stimulation of PI turnover by norepinephrine or carbamylcholine is not affected by the growth conditions. The specific relation between activation of mGluR$_{PIS}$ and the formation and maintenance of the typical shape of astrocytes in chemically defined medium remains to be established.

9. Conclusions

The study of native mGluR$_{PIS}$ expressed in brain slices or cell cultures has revealed that these receptors undergo plastic changes during postnatal development as well as in adult life. The transient rise in receptor activity observed at the earlier stages of development suggests that mGluR$_{PIS}$ regulate developmental plasticity and are important in the transition among neuronal proliferation, differentiation, and apoptotic death, which can be viewed as a specialized phenotype of the cell. Based on results obtained in cultured neurons, we speculate that mGluR$_{PI}$ activation serves to provide a general trophic input that will prevent apoptotic death in developing neurons. When this protective system breaks down, i.e., when mGluR$_{PIS}$ become responsive to homologous desensitization, only those neurons that are innervated by joined synaptic inputs (necessary to activate NMDA receptors) or are targets for specific trophic factors can be rescued from programmed cell death. If the drop in mGluR$_{PI}$ activity will give access to apoptotic death, then one expects that the reverse (i.e., an increase in receptor activity) occurs in response to denervation, when neurons build a defense against the lack of trophic inputs. This is indeed the case, as shown in a variety of models, including chemical or mechanical deafferentation or hypoxic-ischemic neuronal damage. Although mGluR$_{PI}$s are downregulated in adult life, their activity may increase during the learning process, and this will provide an important mechanism that enables various stages of learning, as well as the transition between short-term and long-term memory. Finally, mGluR$_{PI}$s are also expressed in astrocytes, where they appear to regulate morphocytogenesis and their activation increases gene expression. To what extent astrocytes contribute to the dynamic changes in mGluR$_{PI}$ activity observed during postnatal development or in adult life remains to be established.

References

Abe, T., Sugihara, H., Nawa, H., Shigemoto, R., and Nakanishi, S. (1992) Molecular characterization of a novel metabotropic glutamate receptor mGluR5 coupled to inositol phosphate/Ca^{2+} signal transduction. *J. Biol. Chem.* **267**, 13,361.

Akijama, K., Norihito, Y., and Mitsumoto, S. (1987) Increase in ibotenate-stimulated phosphatidylinositol hydrolysis in slices of the amygdala/pyriform cortex and hippocampus of rats by amygdala kindling. *Exp. Neurol.* **98**, 499–508.

Aniksztein, L., Bregestovski, P., and Ben-Ari, Y. (1991) Selective activation of quisqualate metabotropic receptor potentiates NMDA but not AMPA responses. *Eur. J. Pharmacol.* **205**, 327,328.

Aronica, E., Frey, U., Wagner, M., Schroeder, H., Krug, M., Ruthrich, H., Catania, M. V., Nicoletti, F., and Reymann, K. G. (1991) Enhanced sensitivity of "metabotropic" glutamate receptors after induction of long-term potentiation in rat hippocampus. *J. Neurochem.* **51**, 725–729.

Aronica, E., Condorelli, D. F., Nicoletti, F., Dell'Albani, P., Amico, C., and Balzs, R. (1993a) Metabotropic glutamate receptors in cultured cerebellar granule cells: developmental profile. *J. Neurochem.* **60**, 559–565.

Aronica, E., Nicoletti, F., Condorelli, D. F., and Balazs, R. (1993b) Pharmacological characterization of metabotropic glutamate receptors in cultured cerebellar granule cells. *Neurochem. Res.* **18**, 605–612.

Aronica, E., Dell'Albani, P., Condorelli, D. F., Nicoletti, F., Hack, N., and Balazs, R. (1993c) Mechanisms underlying developmental changes in the expression of metabotropic glutamate receptors in neuronal culture. *Mol. Pharmacol.* **44,** 981–989.

Balazs, R., Gallo, V., and Kingsbury, A. (1988) Effect of depolarization on the maturation of cerebellar granule cells in culture. *Dev. Brain Res.* **468,** 269–273.

Bashir, Z. I., Sunter, D. C., Watkins, J. C., and Collingridge, G. L. (1993) Metabotropic glutamate receptors contribute to the induction of long-term depression in the CA1 region of the hippocampus. *Eur. J. Pharmacol.* **239,** 265,266.

Behnish, T., Fedorov, K., and Reymann, K. G. (1991) L-2-Amino-3-phosphono-propionate blocks late synaptic long-term potentiation. *NeuroReport* **2,** 386–388.

Ben-Ari, Y. (1993) Metabotropic receptors and synaptic plasticity: a gating mechanism for the induction of LTP. Abstract presented at the International Meeting on Function and Regulation of Metabotropic Glutamate Receptors, September 19–23, Taormina, Italy. *Functional Neurology* (suppl. 4), 10.

Bessho, Y., Nawa, H., and Nakanishi, S. (1993) Glutamate and quisqualate regulate expression of metabotropic glutamate receptor mRNA in cultured cerebellar granule cells. *J. Neurochem.* **60,** 253–259.

Boss, V. and Conn, P. J. (1993) Coupling of metabotropic excitatory amino acid receptors to phospholipase D: a novel pathway for generation of second messenger. Abstract presented at the International Meeting on Function and Regulation of Metabotropic Glutamate Receptors, September 19–23, Taormina, Italy. *Functional Neurology* (suppl. 4), 12.

Cartmell, J., Kemp, J. A., Alexander, S. P. H., Hill, P. H., and Kendall, D. A. (1992) Inhibition of forskolin-stimulated cAMP formation by 1-aminocyclopentane-trans-1,3-dicarboxylate in guinea pig cerebral cortical slices. *J. Neurochem.* **58,** 1964–1966.

Cartmell, J., Kemp, J. A., Alexander, S. P. H., and Kendall, D. A. (1993) Endogenous adenosine regulates the apparent efficacy of 1-amino-cyclopentyl-1S,3R-dicarboxylate inhibition of forskolin-stimulated cyclic AMP accumulation in rat cerebral cortical slices. *J. Neurochem.* **60,** 780–782.

Casabona, G., Genazzani, A. A., Di Stefano, M., Sortino, M. A., and Nicoletti, F. (1992) Developmental changes in the modulation of cyclic AMP formation by the metabotropic glutamate receptor agonist 1S,3R,aminocyclpentane-1,3-dicarboxylic acid in brain slices. *J. Neurochem.* **59,** 1161–1163.

Catania, M. V., Landwehrmeyer, B., Standaert, D., Testa, C., Penney, J. B., and Young, A. B. (1993) Differential expression patterns of metabotropic glutamate receptor mRNAs and binding sites in developing and adult rat brain. Abstract presented at the International Meeting on Function and Regulation of Metabotropic Glutamate Receptors, September 19–23, Taormina, Italy. *Functional Neurology* (suppl. 4), 15.

Cha, J.-H. J., Makowiec, R. L., Penney, J. B., and Young, A. B. (1990) AP3 and L-BMAA displace [^{3}H]glutamate binding to the metabotropic receptor. Abstract 231.19 presented at annual meeting of the American Society for Neuroscience, October 28–November 2, St. Louis, MO.

Chen, C. K., Silverstein, F. S., Fisher, S. K., Statman, D., and Johnston, M. V. (1988) Perinatal hypoxic-ischemic injury enhances quisqualic acid-stimulated phosphoinositide turnover. *J. Neurochem.* **51**, 353–359.

Condorelli, D. F., Ingrao, F., Magri, G., Bruno, V., Nicoletti, F., and Avola, R. (1989a) Activation of excitatory amino acid receptors reduces thymidine incorporation and cell proliferation rate in primary cultures of astrocytes. *Glia* **2**, 67–69.

Condorelli, D. F., Kaczmarek, L., Nicoletti, F., Arcidiacono, A., Dell'Albani, P., Ingrao, F., Magri, G., Malaguarnera, L., Avola, R., Messina, A., and Giuffrida, A. M. (1989b) Induction of protooncogene fos by extracellular signal in primary glial cell cultures. *J. Neurosci. Res.* **23** (2), 234–239.

Condorelli, D. F., Dell'Albani, P., Amico, C., Casabona, G., Genazzani, A. A., Sortino, M. A., and Nicoletti, F. (1992) Developmental profile of metabotropic glutamate receptor mRNA in rat brain. *Mol. Pharm.* **41**, 660–664.

Copani, A., Canonico, P. L., and Nicoletti, F. (1990) Beta-N-methylamino-L-alanine (L-BMAA) is a potent agonist of "metabolotropic" glutamate receptors. *Eur. J. Pharmacol.* **181**, 327–328.

Copani, A., Canonico, P. L., Catania, M. V., Aronica, E., Bruno, V., Ratti, E., van Amsterdam, F. T. M., Gaviraghi, G., and Nicoletti, F. (1991) Interaction between β-N-methylamino-L-alanine and excitatory amino acid receptors in brain slices and neuronal cultures. *Brain Res.* **558**, 79–86.

Dewar, D., Chalmers, D. T., Graham, D. I., and Mc Culloch, J. (1991) Glutamate metabotropic and AMPA binding sites are reduced in Alzheimer's disease: an autoradiographic study of the hippocampus. *Brain Res.* **553**, 58–64.

Dudek, S. M. and Bear, M. F. (1989) A biochemical correlate of the critical period for synaptic modification in the visual cortex. *Science* **246**, 673–675.

Eaton, S. A., Jane, D. E., Jones, P. L. St. J., Porter, R. H. P., Pook, P. C.-K, Sunter, D. C., Udvarhelyi, P. M., Roberts, P. J., Salt, T. E., and Watkins, J. C. (1993) Competitive antagonism at metabotropic glutamate receptors by (S)-4-carboxypheylglycine (CPG) and (RS)-α-methyl-4-carboxyphenylglycine (MCPG). *Eur. J. Pharmacol.* **244**, 195–197.

Favaron, M., Rimland, J. M., and Manev, H. (1992) Depolarization - and agonist-regulated expression of neuronal metabotropic glutamate receptor 1 (mGluR1). *Life Sci.* **50**, 189–194.

Forscher, P. (1989) Calcium and polyphosphoinositide control of cytoskeleton dynamics. *Trends Neurosci.* **12**, 468–474.

Gallo, V., Kingsbury, A., Balazs, R., and Jorgensen, O. S. (1987) The role of depolarization in the survival and differentiation of cerebellar granule cells in culture. *J. Neurosci.* **7**, 2203–2213.

Genazzani, A. A., Casabona, G., L'Episcpo, M. R., Condorelli, D. F., Dell'Albani, P., Shinozaki, H., and Nicoletti, F. (1993) Characterization of metabotropic glutamate receptors negatively linked to adenylyl cyclase in brain slices. *Brain Res.* **622**, 132–138.

Guiramand, J., Sassetti, I., and Recasens, M. (1989) Developmental changes in the chemsensitivity of rat brain synaptoneursomes to excitatory amino acids estimated by inositol phosphate formation. *Int. J. Dev. Neurosci.* **7**, 257–266.

Harvey, J., Frenguelli, B. G., Watkins, J. C., and Collingridge G. L. (1991) The actions of 1S,3R-ACPD, a glutamate metabotropic receptor agonist, in area CA1 of rat hippocampus. *Br. J. Pharmacol.* **104,** C79.

Hayashi, Y., Moueryame, A., Takahashi, T., Ohishi, H., Ojawa-Meyuro, R., Shigemoto, R., Mizuno, M., and Nakamishi, S. (1993) Role of metabotropic glutamate receptor in synaptic modulation in the accessory olfactory bulb. *Nature* **366,** 687–690.

Herrero, I., Miras-Portugal, M. T., and Sanchez-Prieto, J. (1992) Positive feedback of glutamate exocytosis by metabotropic receptor stimulation. *Nature* **360,** 163–166.

Holler, T., Klein, J., and Loffelholz, K. (1993) Glutamate activates phospholipase D in rat hippocampus. Abstract presented at the International Meeting on Function and Regulation of Metabotropic Glutamate Receptors, September 19–23, Taormina, Italy. *Functional Neurology* (suppl. 4), 26.

Iadarola, M. J., Nicoletti, F., Naranjo, J. R., and Costa, E. (1986) Kindling enhances the stimulation of inositol phospholipid hydrolysis elicited by ibotenic acid in rat hippocampus. *Brain Res.* **374,** 174–178.

Iversen, L., Mulvihill, E., Haldeman, B., and Kristensen, P. (1993) Metabotropic glutamate receptor subtype expression pattern in the rat brain analysed by in situ hybridisation. Abstract presented at the International Meeting on Function and Regulation of Metabotropic Glutamate Receptors, September 19–23, Taormina, Italy. *Functional Neurology* (suppl. 4), 27.

Izumi, Y., Clifford, D. B., and Zorumski, C. F. (1991) 2-amino-3-phosphono-propionate blocks the induction and maintenance of long-term potentiation in rat hippocampal slices. *Neurosci. Lett.* **122,** 187–190.

Jensen, A. M. and Chiu, S. Y. (1990) Fluorescence measurement of changes in intra-cellular calcium induced by excitatory amino acids in cultured cortical astrocytes. *J. Neurosci.* **10,** 1165–1175.

Kelso, S. R., Nelson, T. E., and Leonard, J. P. (1992) Protein kinase C-mediated enhancement of NMDA currents by metabotropic glutamate receptors in Xenopus oocytes. *J. Physiol. Lond.* **449,** 705–718.

Klitgaard, H. and Jackson, H. C. (1993) Central administration of L-amino-3-phosphoro-propionate (L-AP3) inhibits audiogenic seizures in DBA/2 mice. Abstract presented at the International Meeting on Function and Regulation of Metabotropic Glutamate Receptors, September 19–23, Taormina, Italy. *Functional Neurology* (suppl. 4), 28.

Lin, W. W., Lee, C. Y., and Chuang, D. M. (1991) Endothelin- and sarafotoxin-induced phosphoinositide hydrolysis in cultured cerebellar granule cells: biochemical and pharmacological characterization. *J. Pharmacol. Exp. Ther.* **257,** 1053–1061.

Littman, L., Munir, M., Flagg, S. D., and Robinson, M. B. (1992) Multiple mechanisms for inhibition of excitatory amino acid receptors coupled to phosphoinositide hydrolysis. *J. Neurochem.,* 59.

Littman, L., Glatt, B. S., Pritchett, D. B., and Robinson, M. B. (1993) Differentiation of metabotropic receptors using L-aspartate-b-hydroxamate (L-AbHA). Abstract presented at the International Meeting on Function and Regulation of Metabotropic Glutamate Receptors, September 19–23, Taormina, Italy. *Functional Neurology* (suppl. 4), 30.

Masu, M., Tanabe, Y., Tsuchida, K., Shigemoto, R., and Nakanishi, S. (1991) Sequence and expression of a metabotropic glutamate receptor. *Nature* **349**, 760–765.

Milani, D., Facci, L., Guidolin, D., Leon, A., and Skaper, S. D. (1989) Activation of polyphosphoinositide metabolism as a signal-transducing system coupled to excitatory amino acid receptors in astroglial cells. *Glia* **2**, 161–169.

Miller, S., Cotman, C. W., and Bridges, R. (1992) 1-aminocyclopentane-trans-1,3-dicarboxylic acid induces glutamine synthetase activity in cultured astrocytes. *J. Neurochem.* **58**, 1967–1970.

Miller, S., Bridges, R. J., and Cotman, C. W. (1993) Regulation of glutamate metabotropic signal transduction in cultured astrocytes. Abstract presented at the International Meeting on Function and Regulation of Metabotropic Glutamate Receptors, September 19–23, Taormina, Italy. *Functional Neurology* (suppl. 4), 59.

Nakagawa, Y., Saito, K., Ishihara, T., Ishida, M., and Shinozaki, H. (1990) (2S,3S,4S)-a-(carboxycyclopropyl)glicine is a novel agonist of metabotropic glutamate receptors. *Eur. J. Pharmacol.* **184**, 205–206.

Nakanishi, S. (1992) Molecular diversity of glutamate receptors and implication for brain function. *Science* **258**, 597–603.

Nicoletti, F., Iadarola, M. J., Wroblewski, J. T., and Costa, E. (1986a) Excitatory amino acid recognition sites coupled with inositol phospholipid metabolism: developmental changes and interaction with a1-adrenoceptors. *Proc. Natl. Acad. Sci. USA* **83**, 1931–1935.

Nicoletti, F., Wroblewski, J. T., Iadarola, M. J., and Costa, E. (1986b) Serine-0-phosphate, an endogenous metabolite, inhibits the stimulation of inositol phospholipid hydrolysis elicited by ibotenic acid in rat hippocampal slices. *Neuropharmacology* **25**, 335–338.

Nicoletti, F., Wroblewski, J. T., Alho, H., Eva, C., Fadda, E., and Costa, E. (1987) Lesions of putative glutamatergic pathways potentiate the increase of inositol phospholipid hydrolysis elicited y excitatory amino acids. *Brain Res.* **436**, 103–109.

Nicoletti, F., Valerio, C., Pellegrino, C., Drago, F., Scapagnini, U., and Canonico, P. L. (1988) Spatial learning potentiates the stimulation of phosphoinositide hydrolysis by excitatory amino acids in rat hippocampal slices. *J. Neurochem.* **51**, 725–729.

Nicoletti, F., Magri, G., Ingrao, F., Bruno, V., Catania, M. V., Dell'Albani, P., Condorelli, D. F., and Avola, R. (1990) Excitatory amino acids stimulate inositol phospholipid hydrolysis and reduces proliferation in cultured astrocytes. *J. Neurochem.* **54**, 771–777.

Nicoletti, F., Casabona, G., Genazzani, A. A., L'Episcopo, M. R., and Shinozaki, H. (1993) (2S, 1'R, 2'R, 3'R)-2-(2,3-dicarboxycyclopropil)glycine enhances quisqualate stimulated inositol phospholipid hydrolysis in hippocampal slices. *Eur. J. Pharmacol. (Mol. Pharmacol. Sec.)* **245**, 297,298.

Ninomiya, H., Taniguchi, T., and Fujiwara, M. (1990) Phosphoinositide breakdown in rat hippocampal slices: sensitivity to glutamate induced by in vitro anoxia. *J. Neurochem.* **55**, 1001–1007.

Ohfune, Y., Shimamoto, K., Ishida, M., and Shinozaki, H. (1993) Synthesis of L-2-(2,3-dicarboxycyclopropil)glycines. Novel conformationally restricted glutamate analog. *Bioorg. Med. Chem. Lett.* **3**, 15.

Pearce, B., Morrow, C., and Murphy, S. (1986) Receptor-mediated inositol phospholipid hydrolysis in astrocytes. *Eur. J. Pharmacol.* **121,** 231–243.

Pellegrini-Giampietro, D. E., Zukin, R. S., Bennett, M. V. L., Cho, S. H., and Pulsinelli, W. A. (1992) Switch in glutamate receptor subunit gene expression in CA1 subfield of hippocampus following global ischemia in rats. *Proc. Natl. Acad. Sci. USA* **89,** 10499–10503.

Pizzi, M., Fallacara, C., Arrighi, V., Memo, M., and Spano, P. F. (1993) Attenuation of excitatory amino acid toxicity by metabotropic glutamate receptor agonists and aniracetam in primary cultures of cerebellar granule cells. *J. Neurochem.* **61,** 683–689.

Porter, R. H., Briggs, R. S., and Roberts, P. J. (1992) L-aspartate-b-hydroxamate exhibits mixed agonist/antagonist activity at the glutamate metabotropic receptor in rat neonatal cerebrocortical slices. *Neurosci. Lett.* **144,** 87–89.

Reymann, K. and Matthies, H. (1989) 2-amino-4-phosphonobutyrate selectively eliminates late phases of long-term potentiation in rat hippocampus. *Neurosci. Lett.* **98,** 166–171.

Riedel, G., Vieweg, S., and Reymann, K. (1993) Tetanus induced long-term potentiation requires activation of metabotropic glutamate receptors in the dentate gyrus in vivo. Abstract presented at the International Meeting on Function and Regulation of Metabotropic Glutamate Receptors, September 19–23, Taormina, Italy. *Functional Neurology* (suppl. 4), 46.

Sacaan, A. I., and Schoepp, D. D. (1992) Activation of hippocamapal metabotropic excitatory amino acid receptors leads to seizure and neuronal damage. *Neurosci. Lett.* **139,** 77–82.

Schoepp, D. D. and Hillman, C. C. (1990) Developmental and pharmacological characterization of quisqalate-, ibotenate and trans-1-amino-1,3-cyclopentanedicarboxylic acid stimulations of phosphoinositide hydrolysis in rat cortical brain slices. *Biogenic Amines* **7,** 331–340.

Schoepp, D. D. and Johnson, B. G. (1988) Excitatory amino acid agonist-antagonist interactions at 2-amino4-phosphonobutyrric acid-sensitive quisqalate receptors coupled to phosphoinositide hydrolysis in slices of rat hippocampus. *J. Neurochem.* **50,** 1605–1613.

Schoepp, D. D. and Johnson, B. G. (1989) Inhibition of excitatory amino acid-stimulated phosphoinositide hydrolysis in the neonatal rat hippocampus by 2-amino-3-phosphonopropionate. *J. Neurochem.* **50,** 1865–1870.

Schoepp, D. D., True, R. A., and Monn, J. A. (1991) Comparison (1S,3R)-1 aminocyclopentane-1,3-dicarboxylic acid (1S,3R-ACPD)- and 1R,3S-ACPD-stimulated brain phosphoinositide hydrolysis. *Eur. J. Pharmacol.* **207,** 351–353.

Schoepp, D. D., Johnson, B. G., and Monn, J. A. (1992) Inhibition of cAMP formation by a selective metabotropic glutamate receptor agonist. *J. Neurochem.* **58,** 1184–1186.

Schoepp, D. D. and Johnson, B. G. (1993) Metabotropic glutamate receptor modulation of cAMP accumulation in the neonatal rat hippocampus. *Neuropharmacology* in press.

Seren, M. S., Aldinio, C., Zanoni, R., Leon, A., and Nicoletti, F. (1989) Stimulation of inositol phospholipid hydrolysis by excitatory amino acids is enhanced in brain slices from vulnerable regions after transient global ischaemia. *J. Neurochem.* **53,** 1700–1705.

Sergueeva, O. A., Fedorov, N. B., and Reymann, K. G. (1993) An antagonist of glutamate metabotropic receptors, (RS)-α-methyl-4-carboxyphenylglycine, prevents the LTP-related increase in postsynaptic AMPA sensitivity in hippocampal slices. *Neuropharmacology* **32 (9),** 933–935.

Sortino, M. A., Nicoletti, F., and Canonico, P. L. (1991) Metabotropic glutamate receptor in rat hypothalamus: characterization and developmental profile. *Dev. Brain. Res.* **61,** 169–172.

Spencer, P. S., Nunn, P. B., Hugon, J., et al. (1987) Guam amyotrophic lateral sclerosis: dementia linked to a plant excitant neurotoxin. *Science* **237,** 517–522.

Thelen, M., Rosen, A., Nairn, A. C., and Aderem, A. (1991) Regulation by phosphorylation of reversible association of a myristoylated protein kinase C substrate with the plasma membrane. *Nature* **351,** 320–322.

Tizzano, J. P., Griffey, K. I., Johnson, J. A., Fix, A. S., Helton, D. R., and Schoepp, D. D. (1993) Metabotropic glutamate receptor activation produces limbic seizures in mice which can be selectively attenuated by L-AP3 and dantrolene. Abstract presented at the International Meeting on Function and Regulation of Metabotropic Glutamate Receptors, September 19-23, Taormina, Italy. *Functional Neurology* (suppl. 4), 55.

Vecil, G. G., Li, P. P., and Warsh, J. J. (1992) Evidence for metabotropic excitatory amino acid receptor heterogeneity: developmental and brain regional studies. *J. Neurochem.* **59,** 252–258.

Young, A. B., Catania, M. V., Dure, L., Hollingsworth, Z., and Penney, J. B. (1993) Metabotropic glutamate receptors in neurodegenerative disorders. Abstract presented at the International Meeting on Function and Regulation of Metabotropic Glutamate Receptors, September 19–23, Taormina, Italy. *Functional Neurology* (suppl. 4), 57,58.

Weiss, S., Schmidt, B. H., Sebben, M., Kemp, D. E., Bockaert, J., and Sladeczek, F. (1988) Neurotransmitter-induced inositol phosphate formation in neurons in primary culture. *J. Neurochem.* **50,** 1425–1433.

Winder, D. G. and Conn, P. J. (1992) Activation of metabotropic glutamate receptors in the hippocampus increases cyclic AMP accumulation. *J. Neurochem.* **59,** 375–378.

Wroblewski, J. T., Ikonomovic, S., Santi, M. R., Wroblewska, B., and Grayson, D. R. (1993) Expression of metabotropic glutamate receptors in cultures of cerebellar granule cells. Abstract presented at the International Meeting on Function and Regulation of Metabotropic Glutamate Receptors, September 19–23, Taormina, Italy. *Functional Neurology* (suppl. 4), 57.

Wyllie, A. H., Kerr, J. F. R., and Currie, A. R. (1980) Cell death: the significance of apoptosis, in *International Review of Citology* 68, Academic, New York, pp. 251–306.

Index

DATE DUE